THE PROSPECTORS' MANUAL

"LUCKY HAMILTON" MAKES A STRIKE.

THE PROSPECTORS' MANUAL

BY

ARTHUR J. BURDICK
Author of "Just Jingles," "Cactus: the Plant of Mystery," etc.

ILLUSTRATED

FACSIMILE REPRODUCTION
Shorey Publications • Seattle

#U-18

Facsimile Reproduction 1970

Eighth Printing
Limited to 800 Copies
November, 1982

Shorey Publications
110 Union St.
Seattle, Wa. 98101

ISBN #0-8466-6018-0

PREFACE.

IN THE Prospector's Manual the writer has aimed to give plain, simple and practical directions for the guidance of those who would seek the treasures Nature has hidden in her bosom. As much as possible technical terms have been avoided. Facts in language understandable by the novice, directions easily comprehended, suggestions from the stand point of actual experiences, are contained in this volume.

Beginning at home, the pathway of the prospector, or would-be prospector, has been prepared and marked out for him all the way along. An effort has been made to anticipate and answer any and all questions that are liable to arise; to tell him just the things he will most wish to know when he gets into the field. Especial attention has been given to the requirements of the novice—to him who has neither technical knowledge nor actual experience to aid him in his research.

Not for the novice alone, however, is this work prepared, but as well for the experienced prospector whose knowledge has been gained in the field. There are few mineralogists engaged in prospecting. The greater number of those actually in the business began as novices and they have but such knowledge as experience has taught them. As a consequence they are familiar with but a limited number of minerals and are competent only in that limited field.

Most prospectors are seekers for gold, silver and copper. They possibly understand all about those minerals and can readily read the signs which indicate their presence. In searching for these they are as blind, perhaps, to the signs which indicate the proximity of other minerals quite as profitable to mine as those they seek, as the would-be amateur prospector. This want of a more general knowledge is responsible for the lack of development of some of our very valuable mineral deposits. Some grand fortunes are awaiting those prospectors who shall open their eyes to all their opportunities.

A majority of prospectors meet with failure in their efforts to locate paying properties. This is not strange, for a majority of those who go out to seek wealth in this manner take into the field with them little save inexperience and a blind faith in "luck." One who cannot read the handwriting of Nature, who does not even know the alphabet of her language, can scarcely hope to read the secret of her treasure record.

This work aims to teach the rudiments of Natures' chirography. To intrepret some of the signs she has set to serve as a guide to her treasure vaults. To bring to the aid of the reader the results of actual experience in the field. To epitomize, for his benefit, written and classified knowledge on mining and minerals, and on the laws of nature and man governing them, that his chance of failure may be lessened and his opportunity for success made proportionately greater.

ARTHUR J. BURDICK

Los Angeles, Cal., 1905.

CONTENTS.

CHAPTER IV.

SILVER, COPPER AND LEAD.

How to know silver—Silver ore and free silver—Wire silver and conditions under which it is found—Silver glance, black silver, horn silver, etc.—Lead and its associations—Forms in which it is found—Copper and signs which indicate its presence—Carbonates, Sulphides, etc.—General characteristics and description of ores—How to test for silver, for copper and for lead.

CHAPTER V.

OTHER VALUABLE MINERALS.

How to know and where to look for Quicksilver, Tin, Nickle, Cobalt, Iron, Coal, Platinum, Osmium, Iridium, Palladium, Tellurium, etc., etc.—Tests of various minerals and metals—Soluble minerals such as saltpeter, borax, soda, etc.—Where to look for them and under what conditions found—Radium and pitchblend.

CHAPTER VI.

GEMS AND HOW TO FIND THEM.

What gems are worth looking for—Origin of gems and where to expect them—Color, hardness, character and value—Specific gravity of principal gems—Diamonds, Emeralds, Kunzite and Tourmaline—Where diamonds are found in the United States.

CHAPTER VII.

HOW TO TEST ROCKS AND ORES.

Simple and inexpensive tests for the prospector in the field—Use of the blow-pipe—Assaying in camp—How the relative values of ores may be determined in the field—Various methods used by prospectors—Preparing samples for the assayer—Tips that should be heeded.

CHAPTER VIII.

MINING LAWS IN UNITED STATES AND MEXICO.

General laws governing prospecting, locating and operating in the United States—Mining districts—Local laws—Size and extent of claims—Mining laws in Mexico—How to get a permit to prospect—Size of claims and number allowed to the individual, etc.

CHAPTER IX.

HOW TO LOCATE, PROVE AND PROTECT A CLAIM.

What must be done in order to locate a claim—How to locate—Form of notice and how to stake a claim—Assessment work—How to prove up—Extent of claims—Placer, lode, and tunnel claims—Mill site claim—Water rights—General information.

CHAPTER X.

MINING MISCELLANY.

General hints and miscellaneous information—Some mistakes that are made—Prospecting at home A few final words—Glossary of terms and Dictionary of Minerals.

READY FOR THE DESERT

CHAPTER I.

THE PROSPECTORS' OUTFIT.

WHAT IS NEEDED FOR A PROSPECTING TRIP—CAMP OUTFIT AND WHAT IT SHOULD COST—PROSPECTING TOOLS, THEIR COST AND THEIR USE—CHEMICALS AND IMPLEMENTS FOR MAKING ORE TESTS—COMMISSARIES, WHAT TO TAKE IN THE FOOD LINE—HINTS FOR THE CAMP.

The first step in any enterprise is an important one. With the prospector it is of great import, for in selecting his outfit he may make or mar his venture. Not only does the success of his venture depend largely upon a judicious selec- of equipment, but his very life may hang upon it.

In outfitting one of two mistakes is apt to be made. One mistake is in taking too little. The other mistake, quite as fatal to the success of the the expedition, is in attempting to take too much. What things to take and how much to take are matters which depend upon circumstances. It makes a great difference, where the prospector is going, as to the character of his outfit. Also the length of his trip and the duration of it are matters which have an important bearing on his selection. If his trip is to be a short one, and not for a limited time, he will require but a small outfit and it may not be necessary to procure a pack animal to transport his belongings. If however, he is going out on a protracted trip, he will want one or more animals to bear his baggage and may desire one for his own transportation as well.

Right here let it be said that for matters of convenience and for personal safety, as well as for the pleasures of companionship and for considerations of economy, it is well, when it can be done, for two or more persons to join issues in extended trips.

In selecting animals for the trip the characteristics of the field to be prospected must be taken into consideration. In the arid and desert regions, which comprise a large part of the mineral territory of Mexico and Southwestern United States, and in extremely rugged and mountainous districts, the burro is far preferable to the horse—in fact is the only animal practical for the purpose.

The horse is not a good climber, being insecure of foot and too clumsy and timid for the business. Again he must feed heartily and drink frequently or he becomes discouraged, fatigued and incapacitated for travel. The burro, on the contrary, will fast long and will go without water thirty six or forty-eight hours without serious inconvenience. He is sure-footed, fearless on dangerous and difficult mountain trails, and will dine heartily and satisfactorily on sage-brush or greasewood, or upon bacon rinds and leavings of the camp. He will consider himself well pastured in localities which would afford no forage whatever to a horse.

If the journey is to be made in a fertile region,

one which will afford forage for the horse, that animal has its advantages. The horse is capable of carrying a larger burden and is much more swift of foot than the burro. Especially if one elects to ride, a horse is preferable, but in regions devoid of water and vegetation succulent in its nature, select the burro by all means.

Regarding taking animals to ride upon, it must be remembered that one has to depend upon the resources of the country for food and water for his animals, it being impracticable to carry a supply for them, and the fewer animals he has the less perplexities will be in store for him. Most prospectors prefer to walk, taking only so many animals as are required to transport the baggage.

It depends upon the character of the animal and the locality in which it is procured as to its cost. In many of the mountain and desert towns good burros may be purchased for from $3 to $5 per head. In the vicinity of the large towns or in such localities as lie adjacent to much prospected fields, one may have to pay two or three times as much. Pack saddles, which are absolutely necessary, will cost about $3.50 for the single cinch or $4.50 for the double cinch, and kyacks or saddle-bags—"elforkes," the Mexicans term them—may be had for about $5 per pair. They are hung from either side, from the pack saddle and serve to hold the small articles needed on the expedition. A lariat is needed for each animal, with which to picket him out nights.

These will be found handy in helping bind the load to the animal during transportation.

The camp kit should consist of knife, fork and spoon, hunting-knife, with which bacon or meat can be cut, tin or granite-ware cup and plate for each individual and two or three extra plates to hold prepared food, coffee pot and a couple of pails holding about three quarts each in which to heat water and stir batter for cakes, and a frying pan. The whole outfit, including hunting knife, can be procured for $2.50 to $3. A Dutch oven may be added for another dollar, but it is heavy and awkward to pack and is not esential.

An important item in the outfit is the canteens. In the desert country the canteen is the most important part of the equipment and plenty of capacity is essential. Get good canteens! A weak strap or rivet, which will let the filled canteen fall and burst, may cost a life. The best make costs but little more than inferior goods. The gallon size in best quality, costs about 65 cents; six-quart size 75 cents; two-gallon size 85 cents In traveling in dry countries skimp on everything if necessary, but not on water. Take an extra canteen or two to have on hand in cases of emergency, such as the playing out of one in use or the necessity of providing an unusual supply of water. Empty they are light. Full they are life savers.

The matter of blankets is important. Two pairs of heavy woolen blankets are none too many for each individual. In warm weather they will be appreciated when placed between the body and the earth, for the ground is far from being a soft bed. Suitable blankets may be had for from $3.50 to $5 per pair.

A canvas sleeping-bag is not an actual necessity, save in the very cold regions, but they have their uses and advantages and if there are transportation facilities might be added to the outfit. A sleeping bag thwarts the persistent wind in its attempts to strip the blanket from the sleeper, and the use of the bag lessens, very materially, the chances of getting rattlers for bedfellows. Also in case of rain, the protection of the canvas proves most welcome. Sleeping bags are obtainable for $5 and upwards.

In regard to wearing apparel, khaki suits can be purchased for about $4.50 or $5 and are appropriate and serviceable. Unless boots are worn leggins are an absolute necessity. One dollar will buy those which are plenty good. Elkskin shoes are very durable and easy upon the feet. Prospectors boots cost from $4 to $8 per pair.

In every camp should be at least one gun—shotgun or rifle, for with one the prospector is often able to provide fresh meat for his larder. It is con-considered desirable, also, as a matter of protection, to carry a revolver. The cost of these weapons will depend mainly upon the style and quality.

Having settled the matter of camp and personal equipment the matter of prospecting tools arises. What most are needed are these: A prospector's hammer, which will cost about 85 cents; pick with handle, $1; long handled spade, 75 cents; hand-axe, 75 cents; horn-spoon, 50 cents to $1; metal spoon, 15 cents; gold pan, 25 to 50 cents; one or two quart iron mortar, $1 or $1.50, according to size; magnet, 25 to 50 cents; compass 50 to 75 cents; pocket glass 75 cents to $1.75—it is best to pay the higher price and get the best and strongest glass; chemicals and, if practical, a small can of blasting powder and a coil of fuse.

The cost of the chemicals will depend entirely upon the nature of the outfit. A small pocket outfit is now put up which consists of what is termed a pocket smelter, costing $1.50, and acids for making tests, $1. The smelter provides for thirty tests. The tests are little briquettes upon which the pulverized ore sample is placed. The briquette is then burned and the residue is pulverized in a tiny mortar which goes with the outfit and the acid test applied.

For assisting in other tests the outfit should contain a quantity of filter paper, a bottle of iodine, a blow-pipe and small spirit lamp, half a dozen candles and a bottle of methylated spirits or alcohol and turpentine mixed at the ratio of ten parts alcohol to one part turpentine. Ordinarily the flame of the candle with the blow-pipe will be sufficient, the spirits being for such tests as require a great degree of heat. This outfit is neither bulky nor heavy. The manner in which tests are made will be explained in subsequent chapters.

The above outfit is a complete one for protracted trips. Many things may be omitted when transportation facilities are limited, or where but a few days are to be given to the trip. Also, if the course of the prospector takes him through a region where settlements are frequent, where he can add to his supplies at will, it will not be necessary to make so elaborate preparations. One must be guided by circumstances and by his judgment in these matters.

One thing should never be omitted from the prospector's outfit, and that is a hypodermic syringe and a quantity of permanganate of potash. In most mining districts rattlesnakes abound and it is not infrequently the case the prospector is bitten. Unless relief it obtained the bite is apt to prove fatal.

The permanganate of potash comes in crystals. The crystals are dissolved in water and the solution is injected in and about the wound as soon as possible after the wound is inflicted. In order that there may be no delay a small quantity of the solution should be prepared and placed in a vial and kept on hand ready for immediate use.

It has been a popular belief that whisky is an antidote for the poison of a rattlesnake. This is not the case. The poison of the serpent acts upon the blood and veins, destroying the blood corpuscles and gnawing at the veins and walls of the heart. Where a sufficient quantity of the poison is injected into the veins, death is certain unless a neutralizing substance is injected into the blood. This is what permanganate of potash is. Whisky is not a neutralizer, but, in cases where a limited quantity of the poison gets into the blood, it stimulates the the heart during the attack of the poison upon it, and tides the patient over until the power of the poison has been expended.

The prospector will find it to his advantage to make up a small—very small—kit of medicines. This should contain a box of pills, a box of carbolic salve, some court plaster, a package of dental cotton, a box of footease or talcum powder and a cake of pure medicated soap of some kind.

Finally comes the question of food. The novice is apt to make the mistake of including in his stock of provisions a number of things which appeal to his appetite but which have small nutritive value. That which is most nutritive and takes the least space is what is wanted.

Self-raising pancake flour, a little common flour, coffee, tea if desired, beans, rice, bacon, salt, pepper and if convenient a small quantity of vegetables may be taken–a few pounds of potatoes and half a dozen onions, for soup purposes–beef extract in tins, and some canned milk if desired. Do not forget to take a few pounds of crushed barley for the burro, to be used in case all other provender fails, and for his encouragement in hard places. A handful or two, occasionally in the morning, will serve to attach him to the camp and, in case he should some night break his tether it may save you many weary miles chase.

Provisions for a month, on lines above indicated, will not cost to exceed $6 or $8 per person.

Packing the outfit, stowing the baggage upon the pack animals, and binding the load are matters in which experience and experiment will prove the best teachers. Care should be taken not to impose too heavy burdens upon the pack animals, for, aside from considerations of mercy, it must be remembered that the success of the expedition depends largely upon them.

A folded blanket should be placed under the pack saddles and the kyacks, or hampers, should be as evenly weighted as possible. After the smaller and heavier articles have been stowed in the kyacks, the extra clothing and blankets, with whatever else will not go into the kyacks, should be made into a long, compact bundle and securely bound upon the top of the saddle. The pick and shovel may be tied upon top of all in such a manner that they may be easily removed without disturbing the pack, in case they should be wanted during the day's march. Canteens of water may be hung from the horns of the saddle on either side.

If insecurely packed or bound, or if the load is not evenly balanced, the jolting, swinging strides of the animal will cause it to work out of place and may slip the saddle on the beast. Practice, however, will soon enable the novice to put a pack on so that it will hang all day.

The seasoned prospector needs no camp hints. He knows the importance of selecting, when possible, sites where water and fuel are obtainable. He understands the importance of sheltered locations, and will, when he can do so, make camp on the lee side of a bluff, rock, thicket or hummock. He has learned when he lies down to sleep at night, to place himself with his feet to the wind, knowing full well that otherwise the playful wind will creep under his blankets and attempt to strip them from him.

He has also learned the necessity of securely packing all the edibles and storing them, with his saddles, straps, shoes and other leather goods where the coyote cannot get to them, for that noisy animal is a veritable thief and while he makes night hideous with his vocal accomplishments, he can, when it is to his interests, keep remarkably quiet and will steal softly into the circle and take things from the midst of the camp, leaping silently over sleepers in order to do so.

The novice learns these things for himself by experience, if not otherwise. It is pleasanter, however, and less expensive, to gain this knowledge in advance through the experiences of others. It is for this reason that these matters are here mentioned.

CHAPTER II.

WHERE TO GO.

WHERE TO GO AND WHERE NOT TO GO—SOMETHING ABOUT MINERAL FIELDS AND WHERE MINERALS MAY BE LOOKED FOR—CHARACTER OF ROCKS RELATIVE TO MINERALS—WHAT MINERALS ARE MOST VALUABLE AND THEIR RELATIVE VALUES—VEIN MINING AND PLACER FIELDS—VARIOUS FORMS OF ORE DEPOSITS—VEINS, LODES, FISSURES, ETC.—FAULTS, SPLITS, FOLDS AND PINCHED OUT VEINS—WHAT COLORS INDICATE—VEGETABLE GUIDES.

The question where to go depends largely upon the object of the prospector. If he be a general seeker for mineral wealth, and is prepared by education and equipment to recognize the various kinds of minerals, stones and gems, and determine their values, then the world is his field, for there are few areas of great extent which do not contain valuable minerals of some kind.

If he has but one, two, or half a dozen minerals in view, or is qualified to judge of but a limited number, then he best seek those regions known to contain those minerals.

There are certain strips of country known to contain certain minerals, to which the term "belt" has been applied. Thus we have our gold belts, silver belts, copper belts, coal belts and oil belts.

There are belts or areas in which precious and semi-precious gems are found and certain other treasures of the earth are so located as to come within the definition of mineral belts.

While discoveries of metals, minerals and gems are frequently made in entirely new fields, it is in the prospecting of the territory within the known districts that the prospector's best chances for success lie.

Right here let it be said that the best place NOT to visit, is a "boom" locality—the place where a rush is in progress, as the result of some real or reputed rich strike. Avoid such places, unless you happen to be in the immediate vicinity when the strike is made and can be early on the field. A very good motto for the prospector is: "Avoid therush."

Gold is to be found in greater or less quantities in nearly every part of the world. There are certain sections where it exists in so bountiful quantities as to bring those localities into prominence, The principal gold sections of the North American Continent are those states traversed by the Rocky Mountains, the Sierra and the Coast Ranges, certain sections of Georgia, North and South Carolina, Mexico, Alaska, and portions of British Columbia.

Other famous gold producing localities are to be found in Australia, Brazil, New Granada, Chili. Peru, Asiatic Russia, Indian Archipelago, Africa and Hungaria. It is stated, also, that the Philippines give rich promise of becoming important gold producing territory.

Silver is found in greater or less quantities in most gold-producing localities and the two metals are often found associated in the same ores, and frequently in connection with copper. Colorado and Nevada are the two most famous silver producing states, but there are several other states in which a great deal of the metal is found, notably Utah, California, Montana and Idaho. Mexico has fabulously rich mines and in several sections of South America are wonderfully rich mines. Siberia, Saxony, and Hungary also produce large quantities of silver.

It is not practical, in a work of this kind, to go into the history of all the numerous minerals which it is profitable to mine, when found in quantities and under favorable conditions. In that great region of the United States west of the Rockies, are innumerable minerals which are worthy the consideration of the prospector. Of these the following are now being produced with profit in various localities:

Antimony, asbestos, asphalt, bituminous rock, borax, cement, chrome, chrysoprase, clay for brick, clay for pottery, coal, copper, fuller's earth, gold, granite, building and statuary, graphite, gypsum, gems of many varities, infusorial earth, iron ore, lead, lime, lithia, macadam, magnesite, manganese, marble, mica, mineral paint, mineral water, natural gas, niter, onyx, petroleum, platinum, pyrites, quartz crystals, quicksilver, rubble, salt, sand for glass, sand, quartz, serpentine, silver, slate, soapstone, sulpher, tin, and others.

Mexico is a prolific field for the special or general prospector. Her principal minerals and their locations are here given.

Gold and silver in the states of Zacatecas, Hidalgo, Guanajuato, San Luis Potosi, Chihuahua, Sonora, Lower California, Guerrera, and Oaxaca.

Platina in small quantities in Vera Cruz and Guerrera.

Copper in Guerro, Michoacan, Guanajuato, Sonora, Lower California and Chihuahua.

Iron in Hidalgo, Guanajuato, Jalisco, Morelos, Vera Cruz, Lower California, Sonora, Oaxaca, and Durango.

Magnetic iron is found in Durango, at Cerro del Mercado, where is the greatest mass known in the world.

Lead in Zacatecas, Sonora, and Oaxaca.

Zinc in Guerrero.

Quicksilver in Michoacan, Zacatecas, San Luis Potosi, Tabasco, and Guanajuato.

Bismuth in Zacatecas.

Marble and onyx in Lower California, Vera Cruz, Oaxaca, Puebla, San Luis Potosi, and Sonora.

Opals in Queretaro and Guerrero.

Natron, (carbonate of soda), Mexico.

Coal in Vera Cruz, Puebla, Oaxaca, Tamaulipas, Coahuilia, Sonora, and Tabasco.

Sulphur is found in the craters of volcanoes in several states.

Petroleum in Vera Cruz, Oaxaca, and Tabasco.

Asphalt in Tamaulipas and Vera Cruz.

Monumental and statuary granite in Oaxaca, Jalisco, Mexico, and Guerrero.

Porphyry in Chihuahua, Hidalgo, Jalisco, Puebla, Queretaro, Zacatecas.

Potter's clay in Jalisco and Puebla.

A little general knowledge of the character of rocks will prove of great assistance to the prospector. For example, he should be able to distinguish between igneous rocks and aqueous rocks, as very different classes of minerals are common to them. Aqueous rocks are water-formed; igneous are fire-formed. One need never look for coal, for example,in volcanic formation. On the other hand, metalliferous veins are not to be sought adjacent to coal beds.

Certain minerals and groups of minerals are associated with certain kinds of rock. Gold is most commonly found in quartz, though there is scarcely a kind of rock known in which gold is not sometimes found.

The minerals most commonly associated with limestone are zinc and lead. In a hornblende formation one may expect chrome iron, asbestos, soapstone, gold, silver and copper. In shale he may look for coal, fire-clay, iron, rock salt, gypsum.

In granite the precious metals, tin and its allied metals, molybdenum, etc. The volcanic regions offer a field for the gem prospector; also the precious metals, sulphur and a variety of other minerals.

Ancient lake beds, or sinks in the hot desert regions are the natural depositories of soluble minerals such as borax, niter, soda, saltpeter and the like.

There are many minerals other than gold, silver, copper, lead, tin, iron, coal and the more common and better known minerals, which are very valuable. A few of these may be mentioned with their approximate values in the American markets:

Asbestos, $20 per ton; asphaltum $15 to $50; borax, 6 cents per pound; bromine, 45 cents pound; chrome ore, $20 ton; cobalt, $1.50 to $2 per pound; copperas, half a cent a pound; graphite, one to 4 cents per pound; magnesite, $7 per ton; manganese, ore, $2 per ton; ozokerite, 6 cents per pound; monazite, $140 ton; saltpeter, 3 cents pound; soapstone, $10 ton; sodium nitrate, ¾ to 2 cents pound; sulphur, $15 to $20 per ton; antimony, 8 cents pound; iridium, $1 per gram; nickle 30 cents pound; platinum, $15 per ounce; quicksilver, $40 per flask of 76½ pounds; zinc, 5 cents pound; barium, $1 per gram; beryllium, $2.50 an ounce; cerium, $1.50 per ounce; didymium, $4 per ounce; erbium, $3 per ounce; lithium nitrate, 50 cents per ounce; niobium, $3.50 per gram; osmium, 75 cents per gram; rhodium, $2.50 per gram; rubidium, $4.50 per gram; thorium nitrate, $7 per pound; yttrium nitrate, $4 ner ounce; zirconium nitrate, $1 per ounce.

The prospector expects to find his wealth in lodes or veins, or in alluvial deposits known as placers. If in the latter he may, perhaps, by means of the pan, the rocker, the long Tom, or the flume, reap early results of his find.

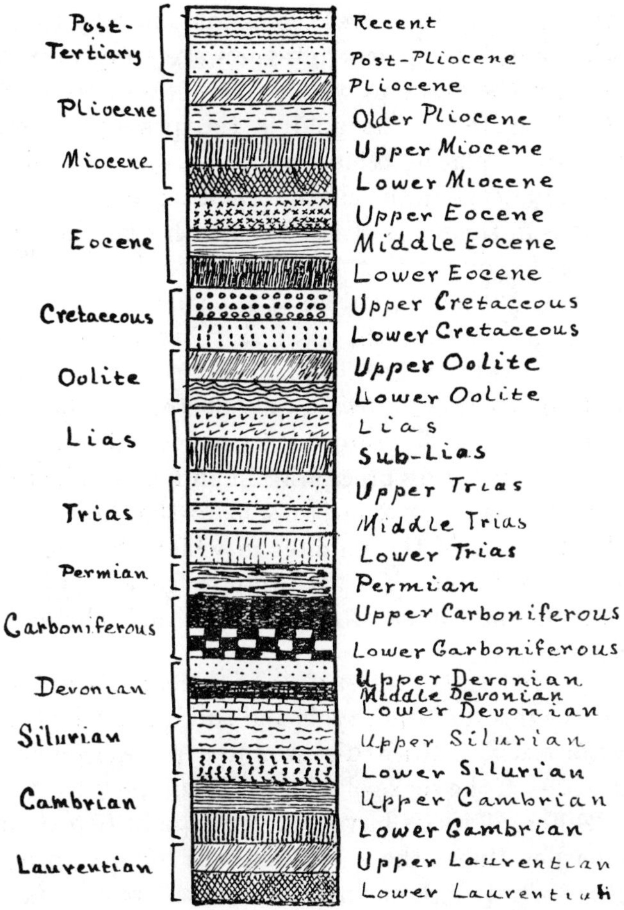

Fig. 1. Fossiliferous Strata as laid down.

Placers serve the prospector in a double capacity: They yield him a profit on their own account and serve, in many instances, to point the way to the mother lodes, for placer minerals are simply washings from the veins in which the metals were originally deposited. As gold is the principal metal taken from placers, this feature of mining will be treated in detail in the chapter relating to gold.

It is in the veins of Mother Earth that the great fortunes lie, and it is in locating these veins, the determining of their extent and the nature and value of their contents that the skill and judgment of the prospector is brought into play.

It matters little to the prospector whether the minerals are injected into the fissures or veins by volcanic pressure or by precipitation and crystallization in water. It is, however, important that he have an understanding of the manner in which the veins or fissures are formed, for this knowledge will aid him in tracing the course of the lodes he may chance to discover.

A great deal of the surface of the earth is broken up as the result of upheavals and sinking down of the earth's crust. Such disturbances produce cracks and seams in the rocks of which it is composed.

Different rocks have different qualities. Some are harder than others. Some are laminated and others are not. Some in breaking will break squarely, others irregularly. A little experience will teach the prospectors the peculiarities of different qualities of rock so that he may judge very aptly whether a

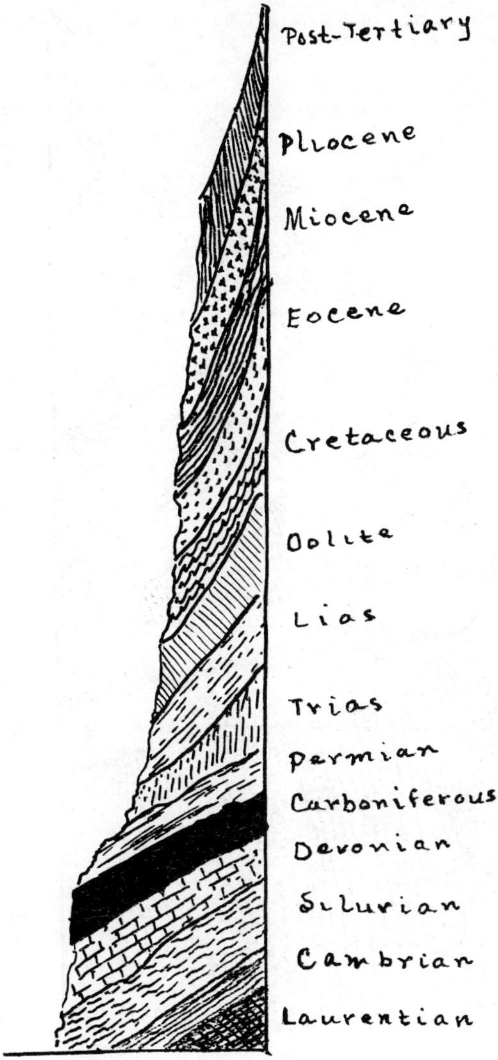

Fig. 2. Fossilferous strata as shown by upheaval.

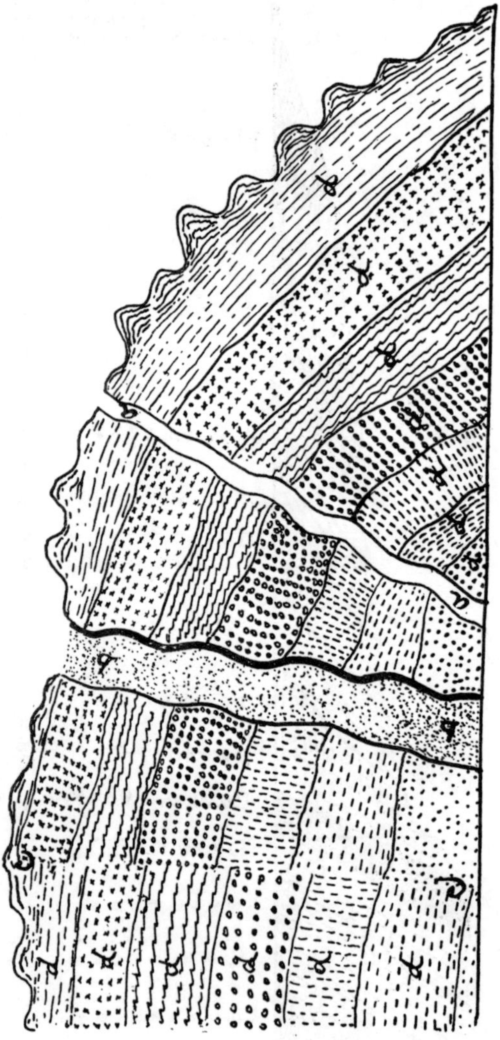

Fig. 3. Sectional view of stratification showing d, d, different strata; a, a, mineral vein; b, b, porphry dike; c, c, fault.

fissure would take a straight or an irregular course through a stratification of that rock.

Upheavals are apt to be followed, at some subsequent period, by reaction or a partial settling of the disturbed masses. This results in a further fracturing of the rocks, in slides producing "faults," in crossing of the veins with others of a later formation, in curving and crooking the old veins, once straight, and in pinching others in places. Pressure upon certain portions of the crust from opposite directions, occasioned by more rapid movement of certain portions of the mass, causes the strata to double up forming "folds." This causes some of the veins and some of the strata to parallel themselves.

Fig. 4. Extreme folding of strata caused by lateral pressure.

As veins are fissures in the rocks which have become filled with metals or minerals, and as these may have been subjected to various disturbances and shiftings subsequent to their being formed, it it evident that a great deal of uncertainity enters into the problem of determining what is behind the outcorp. The lodes may vary greatly in width, dip, direction, and continuity as well as to the nature of their contents from what is indicated at the point of exposure.

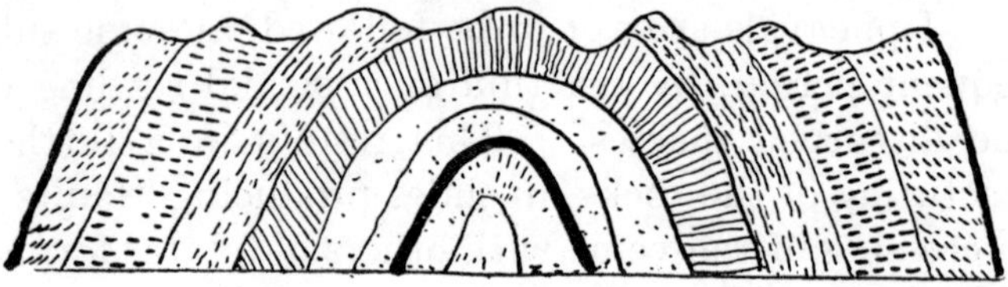

Fig. 7. Fold showing strata running parallel with themselves.

Veins frequently split, branch and run out in a series of small rootlets or stringers. Sometimes they narrow and widen alternately owing to frequent pinches. Again they fault so frequently as to be very difficult to follow. There is, therefore, a great deal of uncertainty attending the following of a mineral vein, but a part of this uncertainty may be removed by learning to read Nature and being able to discover by the external signs what has taken place in the way of earth disturbances, coupling with this understanding the knowledge of the peculiarity of the rocks of that particular region.

The greater number of the outcropping veins are hidden from the eye by an accumulation of soil and geological debris. It is by the use of the pick and and shovel that the prospector comes to these outcropping veins. There are certain surface indications which aid him in determining where these may possibly lie. Some of these will be mentioned in connection with specified minerals. One or two of these pointers given by Nature will not be out of place here.

Some minerals impart a stain to the rocks which contain them, and when the veins outcrop, the exposed surface of the veins are disintregated and disseminated through the soil and the stain is imparted to that as well. Color, therefore, is one of Nature's pointers and a hint the prospector does well to heed.

Black is suggestive of coal, graphite, hematite or black oxide of copper. Brown-black of hauerite. Black is also suggestive of pitch-blend from which radium is obtained.

Blue may mean copper, azurite, lapis lazuli or corelline.

Soil the color of brick dust is the result of iron rust and may indicate iron ore or the decomposition of pyrites. Gold is so frequently found in this form that the prospector looks for placers in this soil and, if he finds color in panning the dirt, for ledges of the mineral in the vicinity.

Green is indicative of malachite, libenthenite, dioplase, abacanite, nickel ocher, texasite, or bromargyrite. Greenish grey, tin pyrites.

Yellow may mean sulphur, copper pyrites, millerite, orpiment, or wulfenite.

Red stains indicate cinnabar, cuprite, pyrargyrite, red ocher or red hematite.

Pink is the color of erythrine, diallogite, rhodonite. In Georgia some of their richest gold mines are in what is known as the "Pink Belt," so called because of the color of the soil.

Vegetation often aids the prospector in locating veins and in tracing their course across the country.

Different substances in the different disintregated rocks attract different plants. Yuccas thrive on granite and quartzite rock; the ocotilla seeks clayey or slatey locations, while the cactus is most frequent on limestone ledges and lava beds or volcanic ash and debris. So marked is the preference of such plants for their habitations that they frequently are found growing in belts and rows as though planted there by man.

It is a popular belief among miners and prospectors that veins are generally richer the deeper down they are followed. They may be or they may not be. There is no logical reason why they should be unless they are now in the position in which they were filled.

Veins and lodes outcrop for one of two reasons. Upheaval has broken the earth, or rocks, tipped the broken edges up and thus exposed them, or the stratification has been cut by water or glaciers, cross-cutting the veins and leaving them exposed. If these veins were filled before these disturbances took place, it may be supposed that the points exposed are fair samples of what is beyond. Of course one must get in on a vein far enough to be beyond the action of the elements upon the minerals, to get a fair sample for assay. This should be considered a fair test of the vein.

When a fissure has been filled after being thrown into its present position, it may be reasonable to suspect that richness would increase with depth if the mineral is deposited by precipitation. There is not enough certainty of such increase to make it worth while to mine a vein that does not show a profit in the ores near the surface, and the prospector will do well to waste little time on such.

CHAPTER III.

GOLD AND HOW TO KNOW IT.

THE HOME OF GOLD–WHERE MOST LIKELY TO BE FOUND–WHERE TO LOOK FOR NUGGETS–FLOAT AND WHAT IT MEANS–HOW TO TRACE "MOTHER LODES"–FREE GOLD–GOLD COMBINED WITH OTHER METALS–FREE-MILLING ORES–SULPHUR ETC.–PYRITES–SIMPLE TESTS FOR GOLD IN ROCKS AND ORES.

Gold is distinctive from other metals in that it is always found in a metallic state, either pure or alloyed with other metals. It occurs in lumps or masses called nuggets; in sheets varying in thickness from a quarter of an inch to that of the thinnest gold leaf; in strings of cube-like crystals from the size of a needle to that of the finest thread, as in what is known as "spider-web" gold; in tiny scales; and in the finest of flour-like dust.

The metals with which gold is found alloyed in nature are platinum, osmium, iridium, tellurium, mercury, silver, copper, iron, etc.;

Gold is distinguishable from other metals, when found in its natural or pure state, by three characteristics: It is malable, soft, being capable of being cut with an ordinary pocket knife, and is not affected by nitric acid. It melts at 2192 degrees F., and volatiizes at a slightly higher temperature. It readily unites with mercury, forming an amalgam from which it may be separated by distillation.

Gold is most commonly associated with quartz, which has been termed the "home of gold," but it is widely distributed through other rocks. It is most frequently found in eruptive rocks in veins, intrusions and overflows in the metamorphic strata, and, as has been said, is as a rule more plentiful in quartz veins traversing the granite and metamorphic rocks, such deposits being termed true fissure veins. The richest portion of the vein is apt to be near one of the contact walls.

Gold is found in a free and pure state in soil and in the beds of streams and in the channels of ancient waterways. Such deposits are termed placers.

Placer mining consists in separating the gold or other metal from the sand, soil or gravel in which it is found, which is simply the debris of auriferous rocks ground fine by the action of heat, cold, water and the like.

The prospector determines the presence of gold in placer deposits by means of his pan. He does this by filling his pan with the earth–with water if obtainable, but without if necessary–and rotating it rapidly causing the heavier particles to sink to the bottom. The lighter materials are removed from the pan by a peculiar flip or twist which can be acquired with a little practice. When the contents of the pan have diminished to a tiny handful, great care is exercised in discarding the worthless portion and the remaining particles are examined to see if they show "color" as the specks of gold are termed.

In prospecting for placers the beds of creeks and streams, the beaches along the course of the streams, dry washes which have once been river beds, and such alluvial deposits as appear to have received the wash of neighboring mountains, should be carefully examined. It should be borne in mind that gold is much heavier than the ordinary rocks and earth, and that it will sink rapidly through the sand and gravel until stopped by some clay, earth or rock which is

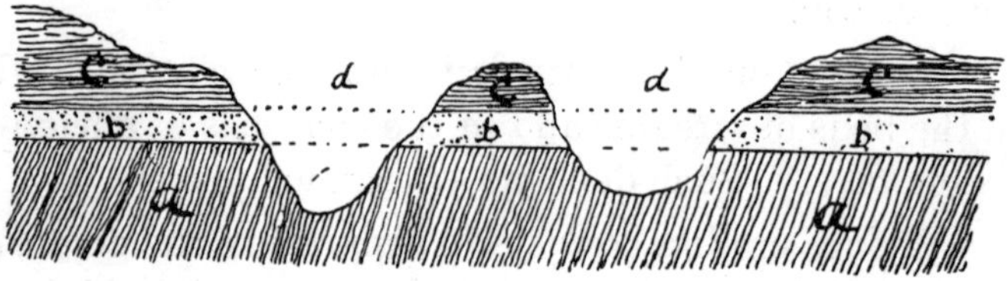

Fig 5. Buried Placer. a, a, bed rock; b, b, b, auriferous gravel in bed of ancient water course; c, c, c, sedimentary rock of recent period; d, d, valleys cut by water or glaciers. Dotted lines show where placer formerly extended.

impervious. It is therefore necessary to sink pits through the debris till such a resistance is encountered. Just above these will lie the pay dirt, if such there be.

In prospecting the stream or the bed of an ancient water course, it is well to remember that gold is less plentiful where the current is swiftest. Behind projecting rocks at the sides of the current, in rock depressions in the bed of the stream, in potholes and eddies, at points where the stream bends sharply–these are the promising places, and here the prospector should make his experiments.

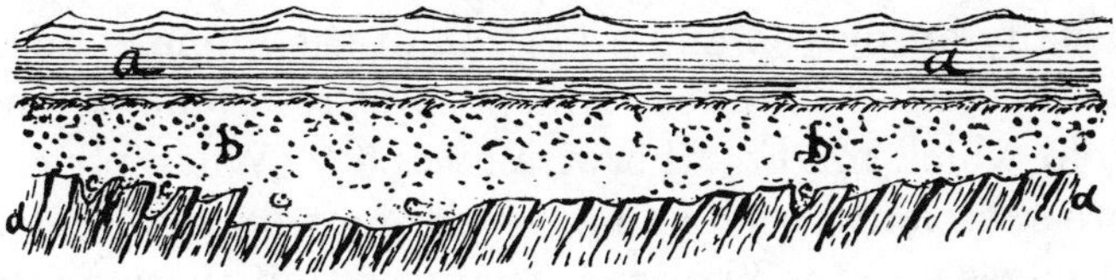

Fig. 6. River Placer. a, a, body of stream; b, b, mud, rocks and coarse material forming bed of stream; c, c, gold-bearing sand and gravel in potholes in d, d, bed rock.

When a rich placer is located the prospector usually wishes to work it at once. He can do this by the old and laborious process of panning the dirt, if no other means are at hand. If he can obtain a little lumber, he may make a rocker or cradle. This is a large wooden trough with cleats nailed across the bottom and rockers underneath into which is shoveled the pay dirt. This is washed by rocking the cradle back and forth, the sand and gravel spilling out at the lower side, the gold being caught and retained by the cleats. If the cleats and box are made perfectly tight and a little mercury be placed in the bottom of the cradle, many fine particles of gold are saved which would otherwise be lost.

When plenty of lumber is to be had and a stream of water can be brought into service, the flume is used and work progresses on a larger scale.

The flume is a box or trough many feet in length so situated that a portion of the stream may be diverted into it. The cleat system is used as in the rocker, only on a more extensive scale, and into the flume, through which the water is constantly running, is shoveled the auriferous earth. Mercury is usually employed and a portion of the flume is lined with woolen blankets which captures some of the gold which might otherwise escape. When placers are sufficiently rich and extensive, machinery, steam dredges and the like are introduced, but these methods are beyond the province of the mere prospector and will not be dealt with in this work.

As has been intimated, the placer should serve as a guide to the mother lode–to the vein or veins from which the gold has been washed. The placer may furnish several valuable hints relative to this.

The nature of the gold found contains hint number one. If the gold taken from the placer is fine, it is pretty safe to conclude that it has traveled a long distance from its original home. If it is found in large lumps, or nuggets, it may be implied, under ordinary circumstances, that the mother lode is near at hand. In other words, the coarser the particles of gold the nearer the lode, as a rule.

There is a hint to be found in the location of the placer. If it is out in a wide plain the hint may be a very faint one, for in such case it may be difficult to determine from which direction the debris has drifted. On the other hand, if the placer is upon a mountain side it is very evident that the mother vein crops out at some point higher up, and to locate it should not be a difficult matter. In narrow canyons and valleys there is a little less certainty, for either wall of the canyon may have contributed the deposit of metal.

There are placer deposits in which the gold and its surrounding matter has been brought hundreds of miles by the glaciers of an ancient period. To trace the source of gold in such deposits would be, of course, impossible.

It has been suggested that the prospector may look for the mother lodes in the vicinity of localities where nuggets are found. It is equally logical to look for nuggets in the vicinity of gold-bearing veins which out-crop. Such prospecting should of course be at points below the out-crop and in line with gravitating forces.

Float is the name given to particles of quartz or vein matter which becomes scattered about the country as the result of the disintregating of out-cropping veins. These are distributed by such forces as water and gravitation. They are valuable aids to the prospector who seeks to locate veins.

Sometimes this float contains particles of free gold, and the prospector is at once assured that the vein is of value, wherever it may be located. Again the pieces show rusty spots and cavities, showing that metal has been oxidized and lost therefrom. This is a promising indication, though not a positive guarantee that the vein is worth finding.

The language of the float is easily read. Small particles having worn or rounded points and edges have traveled far. Large chunks with sharp edges and corners have come but a short distance. Pieces are, as a rule, more numerous in the immediate vicinity of an outcropping vein than farther away.

If, in following a line of float, the prospector comes to a place where it ceases, it is time to turn aside and look for the vein. The finding of it, however, may not be so easy, even now, for it often happens that veins once exposed become covered many feet deep with earth and debris as the result of landslides, floods and other agencies. Men sometimes spend months locating a vein after they have traced it to its immediate locality. It is only by laying bare the rocks that it can be found.

By free gold is meant that which is not confined to veins, such as that found in placers, or to gold which appears in rocks in its pure state, uncombined with other metals. When it is combined with other metals its presence can be determined only by tests, some of which are described below.

Gold is frequently found in pyrites. Pyrites is metallic sulphides whether it be iron, copper, tin, magnetite, or other metal. When other than iron pyrites is referred to, however, it is usual to prefix the metal which enters into the combination, as "copper pyrites" or "tin pyrites."

When gold is associated with pyrites the ores are called sulphides or sulphurets. These ores, if exposed to the elements, or when lying above permanent water levels where they are subjected to alternate wetting and drying, become oxidized, the baser metals rust out, and the gold is left in its pure state. Such ores are caled "free-milling" ores, for the gold has been freed and may be obtained simply by crushing the ores and washing them through a flume or running them over an amalgam plate.

When the prospector finds a piece of rock or ore which he suspects of containing gold, his first test, after a careful examination with the glass, if perchance he may detect free gold, is to pulverize his specimen and "horn" it–that is, wash the powdered rock in his horn spoon in the same manner that he pans dirt. The very finest colors are easily detected in this manner, if the gold is not in combination with another metal.

Assuming that the sample may contain gold in combination with another metal, he will pulverize a piece of the rock and fuse it in his metal spoon, or by burning it upon one of the briquettes described in Chapter I. Then, after pulverizing the residue, he will treat the substance with acid, which releases the gold, if any. The sample is then horned for color.

Here is still another test sometimes employed. The specimen suspected of containing mineral is pulverized in the iron mortar into a fine powder. The powder is then put into the big iron spoon and roasted till it is a cherry red. If the ore fuses it contains metal of some kind. A bit of paper is set fire and dropped into the spoon. If it burns brighter it indicates the presence of nitrates and chlorides. Then the oxidized ore is put into a tin cup and covered with iodine and let stand two or three hours. A piece of filter paper is soaked in the solution and is set on fire. If it gives out a purple color in burning the ore contains gold.

CHAPTER IV.

SILVER, COPPER AND LEAD.

How to know silver—Silver ore and free silver—Wire silver and under what conditions it is found—Silver glance, black silver, horn silver, etc.—Lead and its associations—forms in which it is found—Copper and signs which indicate its presence, Carbonates, sulphides, etc.—General characteristics and description of ores—How to test for silver, for copper and for lead.

Silver occurs occasionally in a pure state in which condition it is easily recognizable because of its silvery white color. Also it is distinguishable from most other metals, which it might resemble, because of its malleability. It may be distinguished from platinum, which it resembles by use of the blowpipe, for silver is fusible while platinum is not. Silver in a pure state sometimes occurs in fibrous particles or threads known as wire silver and sometimes in lumps or nuggets.

Silver is found under other conditions much more frequently than in the pure state. It occurs in silver ores, in ores of lead, and in ores of copper frequently, and with other minerals more rarely.

Argentite is one of the most common of silver ores. Unless newly broken the ore will present a tarnished appearance, but it is easily cut and is then seen to be bright, like newly cut lead. Argentite will melt in the flame of an ordinary candle, without the use of the blowpipe.

Stromeyerine is a deep, lead-grey and is often combined with copper. It is not easy to distinguish save by testing, of which more will be said later.

Stephanite is brittle, otherwise it resembles silver glance.

Pyrargrite and prousite are of a ruby-silver color, the latter being lightest in color.

Horn silver, or kerargyrite, is easily recognized by its resemblance to wax. It is easily cut and quickly tarnishes, becoming a greyish-violet. Bromargyrite is similar to it save that it shows a greenish color.

Iodargyrite is yellow earth, which shows silver under the blowpipe test.

Silver ores are associated with various rocks. In Potasi, South America, and in some portions of California with rhyolite; in different portions of Utah with trachyte; in Honduras with porphry, rhyolite and shale; in various parts of South America with andersite and schists; and in some sections with conglomerates. More frequently the ores are in connection with granite, shale, limestones and dolomites, or with slates.

Lead ores are of several kinds, all easily distinguishable. When heated before the blowpipe all lead ores give a bead of the metal.

Galena, which is a sulphide, is of a metallic-grey color and is about 80 per cent metal. Minium is an oxide found with galena, is distinguishable by its color which varies from orange to red. Cerursite is a carbonate, white to grey in color. Angelesite is n white, grey or black sulphate, very rare. Crocoisite is a chromate, red, which blackens and fuses whea heated. Pyromorphite is yellow, red or green. It swells and changes color when heated. Mimetite is combined with arsenic and is usually covered with a black coating of arsenic. There are about forty other combinations, the metal being associated with antimony, chlorine, and oxygen, sulphur, arsenic, and with phosphoric, antimonious, tungstic, vanadic, chromic, selenious, arsenous and carbonic acids. Some of these combinations furnish cabinet specimens of marvelous beauty, but are otherwise of little value.

Lead ores are found associated most frequently with limestone, granite, gneiss, shales and syenite, but not infrequently with schists, sandstone and quartzite, and occasionally with other rocks.

Copper ore occurs frequently in quartz porphry, but also abundantly in many sedimentary and eruptive rocks. It is a mistake to search for ores in any particular kind of rock, to the exclusion of others, for by so doing valuable deposits may be overlooked.

Copper is easily distinguishable from the conspicuousness of its colors. The metal is associated with a number of other minerals, notably sulphur, antimony, arsenic, iron, zinc, lead, silver, mercury and tin.

The ores, which are numerous, are divided into two general classes: the unoxidized and the oxidized.

In the former class is native copper, which is metallic, ductile and shining; chalcopyrite, yellow and often irridescent, when it is termed "peacock ore;" bornite, purple crystals, very rare; tetrahedrite, commonly known as grey copper, a dark red colored ore, or if particularly rich with zinc, grey; tennanite, crystallized or massive, reddish-grey; euargite, black and very rarely crystallized; bournonite a dark grey ore; stannine, which takes a silvery polish but tarnishes quickly on exposure.

The oxidized ores are melaconite, which is an earthy oxide, staining the fingers black when handled; cuprite, an oxide of bright red color, often colered with malachite; chalcanthite, a blue sulphate soluable in water; azurite, an azure-blue carbonate often in radiated crystallized concretions; malachite, emerald green carbonate, frequently in round protuberances and fibrous; libentheite, a green phosphate, distinguished by small crystals dark upon the surface; atacamite, an olive-green oxychloride, crystallized, granular or fibrous; arseniates of copper are often green and crystallized; chrysocolla is a silicate, green or bluish, and is found in crusts or coatings; dioptase is also a silicate, emerald-green, rarely crystallized.

In the famous copper mines on Lake Superior the metal is found in a pure state in lumps varying in size from an ounce to hundreds of tons. The metal is usually found, however, in the ores known a chalcopyrite, or copper pyrites.

Copper is easily decomposed by the action of the atmosphere and water, it also readily combines with other minerals, which accounts for the multiplicity of its ores. Because of the readiness of the metal to dissipate under the influence of the elements, it is difficult to judge of the richness of a vein by samples taken from near the surface or from such portions of the vein as are exposed to running water or to filterations. By making due allowance for the conditions, however, one may form a reasonable estimate of the value of his find, provided he can arrive at a knowledge of the width, thickness and extent of the vein.

Before proceeding to mention the tests for the various ores of silver, lead and copper, some brief directions for the use of the blowpipe will be necessary.

For the purposes of the prospector the common mouth blowpipe used by jewelers will do. In ordinary tests the flame of a common candle is all that is required. If a great degree of heat is necessary the spirit lamp will be needed.

It will require a little practice, on the part of the novice, to master the blowpipe, but when he has acquired the nack of using it properly, it will be found a simple and efficient aid in determining the nature of the rocks and ores he may collect. It must be borne in mind that it is not necessary to expend a great deal of force in manipulating the pipe, nor does one send the blast through the tube direct from the lungs, but the cheeks should be inflated and the air forced through the tube by gently and steadily contracting the muscles of the cheeks. The tube should be removed from the mouth and shaken, from time to time, to remove the moisture, otherwise the test may be spoiled.

The blowpipe is used for three different purposes: for fusions; for oxidations; and for reductions. Fusion requires the hottest point of the flame. For this purpose the nozzle of the blowpipe should be introduced into the flame, nearly touching the wick, The wick should be freshly trimmed before each test. The blast should be steady and moderately strong.

The inner part of the flame will be a long-pointed bluish cone, brightest near the point. The outer point will be thin, pointed and of a light blue color scarcely visible. The bright point of the inner cone is the hottest point and it is here that the mineral to be fused should be held. The prospector will need a pair of platinum forceps and platinum wire, for making his tests. If it is oxidation that is desired, hold the mineral a little beyond the point of the outer flame, for here it is the oxygen is most plentiful.

It requires a stronger flame and a weaker blast for reducing than in the above described processes. If convenient a smaller tip or aperature to the blowpipe may be used. The force of the blast should be moderate, so as to produce but imperfect combustion. The point of the blowpipe must not penetrate the flame but be held a little above the level of the wick. The flame will then assume the shape of a cone, long, bright, and surrounded by a pale blue flame.

It is well to remember that nearly all sulphides are fusible and some silicates when combined with soda or potash. Zinc forms many infusible compounds. Many oxides are infusible.

In testing rocks and ores with the blowpipe both color and smell are frequently obtained. These have significance and by them the prospector is frequently able to determine the nature of the mineral he has discovered. It is important, therefore, that he familiarize himself with these peculiarities.

Red flame may mean lime, lithia or strontia. Lime gives a yellowish-red flame; lithia a purple-red; and strontia a vivid red.

Yellow flame is peculiar to all soda compounds.

Blue, with a purple border, indicates chloride of copper; greenish-blue flame, bromide of copper.

Violet flame is produced by compounds having potash, provided there is no trace of soda therein.

Green flame suggests baryta, if the flame has a yellowish tint. Copper compounds which do not contain chlorine or bromine give out an emerald-green flame.

The borax test consists in fusing a quantity of borax by taking up on a loop made in the platinum wire a sufficient amount to form a bead. A small quantity of the mineral being tested is fused with the borax, the color being noted to determine the nature of the material. The colors obtained by this means are as follows:

Copper–bead red in reducing flame; blue in oxidizing flame.

Cobalt–deep blue in both processes.

Manganese–colorless in reducing flame; violet in the oxidizing flame.

Nickel–violet in hot oxidizing flame; pale reddish-brown when cold; grey in reducing flame.

Tungstates–violet in reducing flame; colorless in oxidizing flame.

Chromium–flame green in all processes.

Iron–bottle-green in reducing flame; yellow to red in oxidizing flame.

Uranium–yellow in oxidizing flame; green in reducing flame.

Minerals containing arsenic give off a smell of garlic before the blowpipe. Sulphur is easily detected by its fetid fumes. A smell like horseradish is due to selenium.

Some minerals have only to be rubbed, struck with a pick or hammer, or otherwise disturbed to to give off a smell, as for example, asphalt and other bituminous substances, sulphur, and some of itscompounds, and mispickle, which smells of garlic when freshly broken.

Soluble minerals may be determined, oft-times, by their taste. Borax is sweet; niter cooling; alum astringent; sal-amoniac pungent; vitrol acid; sulphate of magnesia bitter.

In silver ores, argentite is most valuable. It fuses readily with or without the acid of the blowpipe, and is detected from grey copper, which it much resembles in appearance, by the fact that it does not give off antimonial fumes nor the sməll of garlic. Iti s much heavier than the copper compound, its specific gravity being 7 and that of grey copper 5.

Stromeyerine is also ordorless but being associated with copper is hard to distinguish without an assay.

Prousite gives off the arsenic smell before the blowpipe; pyrargyrite, fumes af antimony. When crystalized they resemble hematite but they are fusible while the hematite is not. These ores are ruby, and when compact, may be mistaken for cinnabar. The blowpipe will disclose the difference, however, for before the blowpipe, mercury, being volatile, disappears.

All lead ores, when subjected to the blowpipe, give off a metallic bead.

The acid test serves to detect the presence of copper in most of its ores. The test is a simple one. The mineral or ore suspected of containing copper is dissolved in nitric acid and will, if copper be contained therein, deposit the metal in a red metallic state upon polished iron dipped in the solution. The ores that may thus be determined are chalcopyrite or copper pyrites; cubanite; barnhardite; bornite; chalcocite, or vitreous copper; tetrahedrite, or grey copper; melaconite, or black oxide of copper; malachite, or green carbonate of copper; azurite, or blue carbonate of copper.

A few of the ores which prove profitable to mine do not yield to the acid test. Among these are chrysocolla, or copper silicate; atacamite, or chloride of copper; chalcocite, sulphate of copper, or blue-stone. There are about fifty other ores in which copper is an ingredient, but they are of little copper value.

Grey copper yields fumes of antimony before the blowpipe; tennanite gives off the odor of garlic; covellite shows a blue flame; red-ruthite smells of sulphur and is very quick to fuse; bournonite is easily fusible and emits fumes of antimony; azurite gives a blue flame is fusible and when powdered and treated with acid effervesces; malachite burns green, is fusible and effervesces in acid; atacamite colors the flame from the blowpipe blue, is fusible and non-effervescent in acid; libenthenite and arseniate are fusible, soluble, non-effervescent in acid, and the latter smells of garlic in treatment before the blowpipe; chrysocolla and dioptase are green and infusible.

CHAPTER V.

OTHER VALUABLE MINERALS.

HOW TO KNOW AND WHERE TO LOOK FOR QUICKSILVER, TIN, NICKEL, COBALT, IRON, COAL, PLATINUM, OSMIUM, IRIDIUM, PALLADIUM, TELURIUM, ETC.—TESTS OF VARIOUS MINERALS AND METALS—SOLUBLE MINERALS SUCH AS SALTPETER, BORAX, SODA, ETC.—WHERE TO LOOK FOR THEM AND UNDER WHAT CONDITIONS FOUND—RADIUM AND PITCHBLEND.

Quicksilver, or mercury, is very different in its characteristics from all other metals. It occurs in nature principally as a sulphide, though it is occasionally compounded with other minerals. The known ores are:

Amalgam, a compound of silver and mercury.

Ammiolite, an earthy powder, deep red or scarlet, containing mercury, antimony, copper, sulphur and iron.

Barcenite, similar to ammiolite, but contains no copper.

Calomel, or horn quicksilver, light-yellowish, or grey.

Cinnabar, the principal mercurial ore, is bright red to brownish-red and brownish-black, streaks deep red, and before the blowpipe, it volatizes, giving off a strong sulphurous odor. If mixed with dry carbonate of soda and heated over a candle or the flame of a lamp in the iron spoon, it gives off mercurial vapor which may be condensed upon a gold coin held half an inch above the mixture. This is detected by the appearance of the coin, which becomes whitish when rubbed, showing brilliant amalgamation. The mercury may be removed from the coin by heating. The test will show a very slight trace of mercury in ore, so small as one per cent. being disclosed.

Hepatic cinnabar, or liver ore, contains carbon and clay. The ore is a dark, reddish-brown, liver color, and the blowpipe flame shows the same color.

Metacinnabarite, or black sulphide of mercury, resembles graphite in appearance.

Cocainite, or iodide of mercury, is a very rare ore, red, yellow, green, or greenish-grey in color.

Coloradoite, telluride of mercury, is a rare ore, also, and is a greyish-black.

Guadalcazarite, sulphide of mercury, is very similar to metacinnabarite. It is quite rare.

Lehrbachite, is a combination of selenide and mercury. Rare.

Livingstonite, is a greyish-black combination of mercury sulphide and antimonium. Rare.

Magnolite, mercurial tellurate. Rare.

Native mercury, found in many quicksilver mines.

Onofrite, a sulphide containing selenium. Rare.

Tiemanite, dark steel-grey, selenide of mercury. Rare.

Tocornalite, iodide of silver and mercury. Pale yellow, granular and massive. Rare.

Idrialite and aragotite sometimes contain mercury.

Mercury is invariably found in highly metamorphosed rocks–that is, rocks which have been subjected to plutonic or igneous action–and nearly always is found associated with recent eruptive rocks and with hot springs.

It is a peculiar fact that all quicksilver mines yet operated have proven shallow, the richness in mercury decreasing with depth. The finding of native quicksilver is an indication of the impoverishment of the vein.

The localities most productive in mercury in the past are Almaden, Spain, Illyria, Europe, Idria, Austria, Steamboat Springs, Nevada, limited sections in Colorado, and California in various sections, notably New Almaden and New Indria. The California belt extends about 300 miles along the Coast Range.

The cost of reduction of the ores of cinnabar is comparatively small and ores containing a very small percentage of mercury are profitable to work.

Tin is found in abundance in but about half a dozen localities in the world. Three of these are in the United States, ie. Maine, the Black Hills of South Dakota, and in California. The other tin producing sections are Cornwall, England, Banca, East India Islands, and the Malay peninsula. In every instance the veins have been found in quartzose rocks, granite or schists. It is associated only with those rocks which countain white mica. So marked is this peculiarity that granites containing an abundance of white mica have been termed "tin granites." Other minerals commonly associated with tin are topaz, tourmaline, fluorspar and apatite.

The ores of tin are few. Cassiterite, or tin oxide, is practically the only one worth considering. It occurs crystallized and massive; has an adamantine luster; is usually brown or black, though sometimes red, grey, white, or yellow. Its best evidence is its weight, the ore showing little other indications of containing metal. It is brittle, yields but slightly to acids, streaks white, greyish or brownish.

Stannite, or tin sulphide, is steel-grey, when pure. Wood-tin and stream tin, are simply particles of the metal distributed by water after the decompotition of the ores have taken place.

The ores of nickle and cobalt are frequently associated, and both are often found in combination with copper ores. The ores are of two classes--those in which the metals are combined with sulphur or arsenic, or both; and those in which the metals are oxidized. The veins occur in serpentine rocks or in gneiss, hornblend, schist, quartzite, granite, etc.

These minerals are not so readily detected as are some, unless the outcroppings are thoroughly oxidized. Then they may be recognized by their stain. The arseniate of cobalt is a peach-blossom-pink, and it is called "cobalt bloom." The arseniate of nickle is apple-green and is termed "annabergite,"

Both stains may be seen in the same ore, when the two metals are combined.

Except in those ores where nickle and cobalt are combined, the chief source of nickle is in the ore known as pyrrhotite, or magnetic iron pyrites. This is a bronze-colored ore, easily tarnishing, brittle, magnetic, and streaks dark-greyish to black, giving a similar color in the flame of the blow-pipe.

Niccolite, nickeline, or copper nickle, also called arsenical nickle, has a metallic luster, is a pale copper red color and tarnishes. Is not magnetic.

Gersdorffite, or nickle glance, varies in color from silver-white to steel-grey. It usually contains nickle, arsenic and sulphur, but may have, in addition to those minerals, iron or cobalt.

Genthite, or silicate of nickle, is a pale apple-green color which shows a greenish-white tint in the flame.

Mispickel, or arsenical pyrite, is the chief source of both cobalt and arsenic. It frequently carries gold as well. Its distinguishing features are its silvery-white color, metallic luster, is brittle and streaks dark greyish-black.

Smaltite, or cobalt arsenide, is a tin-white ore, when massive, sometimes irridescent, brittle, usually carries nickle.

Cobaltite, or cobalt glance, has a metallic luster, is silvery-white with a reddish tinge or steel-grey with a violet tinge, or if it contains a large percentage of iron, greyish-black. It is brittle. Streaks greyish-black. The above are the principal nickle and cobalt ores.

The blowpipe test will disclose certain things when both metals are combined. First, if but little iron is contained in the ore the bead will be blue. Second, if there is much iron it will at first be a bottle-green, changing to green of a bluish cast. Third, the bead being removed from the platinum wire and heated in a borax bead becomes blue.

Iron enters into the composition of many ores. Here will be mentioned only those yielding iron in paying quantities. These are hematite, goethite, limonite, magnetite and siderite.

Iron is found in a pure state very rarely. It occasionally occurs in this form in lava beds, having been smelted by nature and preserved from decomposition by being incrusted in the lava. In the mines at Negaunee, Michigan, the writer has seen ore 95 per cent pure. As a rule the ores run much under this percentage. Ores yielding 60 to 70 per cent of metal are rich ores.

Magnetite is an oxide of iron, containing 70 per cent of metal, streaks cherry-red.

Limonite, or brown hematite, is what is commonly called "bog ore." It is a hydrous oxide, ochre-yellow color, silky luster, streaks yellowish-brown. It contains about 62 per cent of metal. These ores furnish the bulk of iron of commerce.

Goethite is similar to limonite, the only difference being in the percentage of water in the ores.

The former contains about 16 per cent and the latter from 8 to 10 per cent.

Siderite, or spathic iron, is a carbonate. It occurs crystallized or massive; is brittle; color, ash-grey to greenish-grey, rarely green; streaks white; contains about 62 per cent iron.

Coal is of vegetable origin. It is found imbedded between layers of sedimentary rocks–that is aqueous formed rocks–nearly always in alternating beds of shale and sandstone, sometimes interspersed with beds of fire clay and ironstone.

Originally the stratification was level, but more recent earth shiftings may have caused the strata to dip. Sometimes this upheaval or down-sinking has been sufficiently marked to break up the strata and tilt the beds to an angle, causing them to outcrop. Again water has cut through the strata, forming valleys and exposing the strata and seams of coal.

These outcroppings are the prospector's opportunity. It is only at such points that he may hope to discover, by ordinary methods of prospecting, the coal deposits, for in their natural position they lie buried far beyond his reach.

His chief concern, when he has discovered one of these outcrops, is to learn, first its quality and richness, its depth, width and extent, if the seam dips, and at what angle, the direction in which it extends, and whether the seam be a continuous one or if it pinches out, faults, or is otherwise disturbed.

Without the aid of the drill he can learn these things only by a careful study of the geological con-

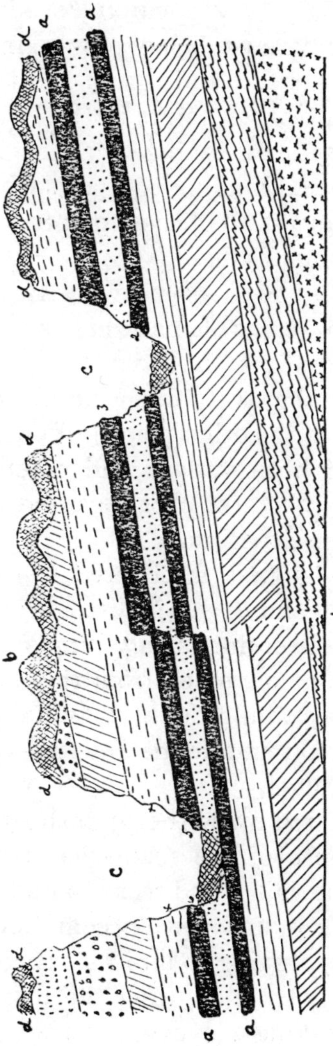

Fig. 8. Showing Faulted Coal Seams. a, a, a, a, seams of coal; b, b, fault; c, c, water cut valleys exposing seams; d, d, d, d, stratum of soil upon surface.

ditions prevailing. By a careful study of the stratification above and below the seam or seams, by observing the trend and dip of the rocks of the region, by seeking other points in the locality where strata are exposed, by taking levels or altitudes at different points and making comparisons, he is able to determine many or all of these points.

In the accompanying diagram (Fig. 8) d, d, d, d, d, d, represents the country level with soil deposit; a, a, a, a, are two seams of coal which are seen, in the sectional view, to dip toward the west; c, c, are valleys cut through the strata exposing in one instance both seams and at the other point but one.

By visiting both valleys and carefully noting conditions the prospector obtains a reasonable comprehension of the situation.

Supposing he first discovered the outcrop at points marked 1 and 2. His discovery was doubtless made by observing small lumps of the coal which had become detached from the seams. The seams themselves would doubtlessly have been covered with loose soil and debris and he exposed them with pick and shovel.

Having cleared away the rubbish sufficiently to disclose the depth of the seams he crosses to the other side of the valley and seeks to find them there. Here he succeeds in uncovering them, but finds them at a lower level, proving to him that there is a dip to the westward. By measuring the distance across the valley and taking a level to ascertain the amount of drop in that distance, he can compute the degree of dip of the seams. Then he passes on to the netx valley, following, if convenient, the angle which the veins are supposed to follow. Arrived there he may reasonably expect to find the seams, or the upper seam, outcropping at x, x. As they do not show there he continues prospecting till he uncovers the outcrop of the one seam at 5 and 6. By the character of the rocks lying above and below the seam he identifies it as the upper seam of his discovery in the other valley.

This seam having dropped several feet below its angle of incline, it is evident that it has faulted somewhere between the two valleys. These faults, their frequency and depth, are problems which effect the cost or practicability of mining the coal.

Of course it is not always that the prospector finds the seams thus conveniently exposed at so many points. However, by the study of such rocks as are exposed in valleys and cuts, he may deduce what has happened to the seam he has discovered.

There are several grades of coal, but commercial coal is classed under two general heads: anthracite or hard coal; and bituminous, or soft coal. These are so well known as to require no description.

Platinum is one of the noble metals. It is one of the least fusible of metals and is very valuable. It occurs in alluvial deposits in small grains, generally with osmium, iridium, and palladium, and frequently in connection with gold. It is found in some parts of California in sufficient quantities to make it profitable to mine. It is a whitish-grey metal, shining, and malleable. It is soluble only when heated in nitro-muriatic acid.

Osmium and iridium combine and form a tin-white or lead-grey colored mineral known as iridosmine. It is malleable only with great difficulty.

Palladium is composed of palladium, irridum and platinum. It is of a whitish-grey color, is ductile and malleable.

Tellurium combines with several minerals. It is the only metal which has been found in nature in actual chemical combination with gold. It occurs very rarely in a pure state. When so found it it of a tin-white color, easily fusible, burns with a greenish flame. Its principal ores are nagygite, a fusible, lead-grey compound, which gives a blue flame before the blowpipe; hetsite, lead-grey and malleable; petzite, iron-black, rare; sylvanite, steel-grey, to silver-white, gives a greenish-blue flame before the blowpipe; calverite, yellowish-grey, giving a bluish-green flame.

Tellurium is chiefly found in this country in Colorado.

The field of the prospector for soluble minerals is limited, for with the exception of salt, this class of minerals is mostly confined to the arid and desert regions, where they are concentrated by evaporation.

Borax occurs in salts and in crustations and in fibrous masses called "cotton balls" in and about dry lakes or boreac lakes and springs. It is usually white, sometimes grey, bluish or greenish. It is brittle, sweetish to the taste, soluble, and imparts a green color to the flame before the bowpipe.

Borate, ulexite, or hayesine, is white, fibrous, tasteless and occurs in balls from the size of an apple to those the size of a bushel basket. This is the variety known as "cotton balls." It is pure and valuable as compared with other forms of crude borax.

Nitrate of potash, niter or saltpeter, is white, has a salty and cooling taste, is soluble, and colors the flame violet when burned. It is found only in hot, dry climates. Wonderful deposits of this mineral have been found in Death Valley, California. It is found upon the surface of the ground or in the stratum of soil which lies at the surface or upon the walls and floors of caverns.

Nitrate of soda, or Chili-saltpeter, has properties similar to the above. It gives a yellowish-flame before the blowpipe. It is found chiefly in South America, though not unknown in California.

Carbonate of soda is so well known as scarcely to need mention. As found in nature it is grey or yellowish-white, alkaline in taste, soluble in water, and effervesces in acid. It is found in large quantities in Nevada and California, particularly in Death Valley.

Sulphate of soda is whitish or brownish in color, soluble in water, is found in dry lakes, notably in Colorado, New Mexico and Wyoming. It is used in the manufacture of gunpounder, glass and soda.

Not long ago a new mineral was discovered to which was given the name of radium, because of its remarkable radio-activity. Comparatively little is yet known of the substance, relative to its mineral history and associations, but scientists are investigating it and are making discoveries regarding it. It is almost priceless in value, being valued at hundreds of dollars per ounce.

Relative to the home of radium, it has been discovered that it is associated, in some instances, at least, with pitchblend or uranium. This is a black, pitchy mineral, the chief constituents of which are the metal uranium, oxygen, lead, iron, etc. It has been used in the manufacture of uranium soda, which furnishes the orange color used in staining glass and in painting china and the like, and it has market value for these purposes. In view of its connection with radium it will be of still greater interest to the prospector.

An ore or composition recently taken from a mine in Mariposa county, California, containing radium, has been called by Prof. Gage, "uranio sulphate ore." He says that it contains uranium, tellurium, gold, copper, silver and radium. He describes the substance as having the faculty of giving off a counter potential energy in the form of an electric current, and is found in the ore after having been treated with chemicals and electricity.

The ore is described as being of a peculiar evasive color, now silver, now grey, now almost black, according to the light in which it is held.

When scratched with a hard substance in the dark it emits a phosphorescent flash. It is stated that in mining the ore it gives off a brilliant flash of light with every stroke of the pick.

Another discovery recently made, relative to this new mineral, is that the mud and soft rock about certain hot springs contain radium. The theory now advanced being that the heat of the waters is caused by their running over deposits of radium. This should be a hint to the prospector not to neglect the rocks of localities about these heated waters.

CHAPTER VI.

GEMS AND HOW TO FIND THEM

WHAT GEMS ARE WORTH LOOKING FOR–ORIGIN OF GEMS AND WHERE TO EXPECT THEM–COLOR, HARDNESS, CHARACTER AND VALUE–SPECIFIC GRAVITY OF PRINCIPAL GEMS–DIAMONDS, EMERALDS, KUNZITE AND TOURMALINE–WHERE DIAMONDS ARE FOUND IN THE UNITED STATES.

There are a number of precious stones and gems which are worth the prospector's while to locate, if he can do so. There are still greater numbers which have more or less commercial value which are, nevertheless, so common as to be of little object to the prospector and miner, unless conditions for their production are extremely favorable.

A few stones stand out far ahead of all others in their value and desirability. Three stones especcially valuable and in active demand at all times are the diamond, the ruby, (red corundum) and the emerald. Good stones of the emerald and ruby are quite as valuable, today, as the diamond.

Tourmalines, rubellites, which are simply red tourmalines, opals, if of good quality, amethyst, topaz, turquoise, beryls, aquamarines and some others are marketable at lesser values.

Precious stones and gems are classed, relative to their degree of hardness, under two heads: Those which are harder than quartz and those which are not. Under the former head are:

Diamond, corundum, chrysoberyl, spinel, topaz, emerald, beryl, phenakite, gircon, dichroite, tourmaline, kunzite, garnet.

Under the head of those which quartz will scratch are:

Amethyst, chalcedony, agate, onyx, sardonyx, carnelian, chrysoprase, plasma, bloodstone, cat's eye, opal, orthoclase, oligoclase, labradorite, olivine, epidote, kyanite, andalusite, turquoise, ultramarine, idocrase, oxinite.

Most gems are crystallizations of materials which have been in a molten state, and are to be expected in volcanic regions, or are the crystallized substances originally in solution and are native in sedimentary rocks. Gems of igenous origin are frequently found, however, in alluvial deposits, having been scattered by floods, rivers and glaciers.

For convenience of the prospector in testing his finds, is given the following table of

SPECIFIC GRAVITY OF GEMS.

Agate	2.5 to 2.8
Amethyst	2.5 to 2.8
Andalusite	3.1 to 3.2
Beryl	2.6 to 2.8
Bloodstone	2.5 to 2.8
Carnelian	2.5 to 2.8
Cat's eye	2.5 to 2.8
Chalcedony	2.5 to 2.8
Chrysoberyl	3.7 to 3.8

Chrysolite	3.3 to 3.5
Chrysoprase	2.5 to 2.8
Citrine	2.5 to 2.8
Cordierite	2.5 to 2.7
Corundum	3.9 to 4.2
Crocidolite	3.2 to 3.3
Diamond	3.5 to 3.6
Dichroite	2.5 to 2.7
Emerald	2.6 to 2.8
Epidote	3.2 to 3.5
Garnet	3.5 to 4.3
Heliotrope	2.5 to 2.8
Kunzite	3.1 to 3.2
Labrodorite	2.68 to 2.74
Lapis lazuli	2.38 to 2.42
Oligoclase	2.64 to 2.68
Olivine	3.3 to 3.5
Onyx	2.5 to 2.8
Opal	2.0 to 2.2
Orthoclase	2.53 to 2.58
Phenakite	2.97
Plasma	2.5 to 2.8
Ruby	3.9 to 4.2
Sapphire	3.9 to 4.2
Sardonyx	2.5 to 2.8
Spinel	3.4 to 4.1
Topaz	3.0 to 3.6
Tourmaline	3.0 to 3.6
Turquoise	2.6 to 2.8
Vesuvianite	3.3 to 4.0
Zircon	4.0 to 4.7

While most gems have a distinguishing color, or are distinguished, as is the diamond, for being colorless, nearly all of them display exceptions to the rule. Diamonds are not only colorless, but they are sometimes found of a pale yellow color, very rarely red, and equally rarely blue. A blue diamond, if flawless, will command two or three times the price of the ordinary stone. A few diamonds of a green tint have been discovered.

Corundum is colorless, red, deep yellow, emerald blue, and rarely violet.

Chrisoberyl is green in various shades.

Spinel is ruby, red, various tints of green, and rarely blue.

Topaz is colorless, deep yellow, pale yellow, pale blue.

Emerald shows various tints of green, emerald pale blue.

Beryl, various shades of green, pale blue, very rarely rose-color.

Zircon, colorless, red-brown, and brown-yellow.

Corderite, indigo, brown.

Phenakite, colorless, deep-yellow, pale-yellow, brown, yellow.

Kunzite, colorless, lilac, greenish tints.

Tourmaline, colorless, pink, red, green, blue, black.

Diamonds are found in various parts of the world, but nearly all the stones used for ornamental purposes come from Brazil and from South Africa. There is no reason to doubt, however, the existence of rich fields in other parts of the world, even in our United States. In fact there are indications that such may be the case.

Stones of value have been found occasionally in Georgia, North Carolina, California, and there are proofs of several having been found in Mexico of excellent quality and size. It is not improbable that extensive deposits of these precious stones will yet be discovered here at home.

Some few hundred thousand years ago, more or less, that portion of the American continent now known as California lay at the bottom of the ocean. Then old Pluto punched up the fires of the under world and the submerged land got its back up and shook off the waters of the ocean, leaving the ground plan of the Golden State high and dry. Old Pluto, however, kept right on shoveling fuel into his underground furnace, with the result that this edge of the Western Continent began to blister and break out into eruptions, and there were hot times in the vicinity for a while.

As there was no one here, at that time, to get the benefit of this display, old Pluto finally became tired of playing to an empty house and turned off the fireworks. All up and down the state–and the Pacific coast, for that matter–are the marks of his little tragedy act. The story is read in the lava, trap and ashes, and in the silent craters which once poured them out.

These marks are not peculiar to this section. In various parts of the world are records of Pluto's picnics. In several parts of the world, too, it has been ascertained that something more valuable than ashes and lava were left as mementos of the warm season. The diamond mines of Kimberly, South Africa are in the bowl or crater of an extinct volcano. The diamonds, it is believed, were formed in crystallizations about the walls or sides of the crater.

What has happened in craters in other parts of the world happened to a certain extent in some of the craters in California. To just what extent it remains for the enterprising prospector to demonstrate. It may be that we have no very valuable diamond fields in this country, and it is possible that the wealth of a Kimberly is awaiting the fortunate discoverer.

About thirty years ago, Chris. Wisner, a California prospector, found a score of diamonds about a mile north of Volcano, in Amador county. He was seeking gold and was running a tunnel through a gravel formation, which lay beneath forty feet of lava. All the stones were found in the pan while testing the gravel for gold and within the space of ten or twelve weeks, the length of time he was engaged in the prospect. Some of these stones were a little off color, but the most of them were clear, and they were genuine diamonds. He made no effort to discover the gems, not knowing their value, merely saving such as happened to intrude themselves upon his notice. In all probability he threw away more than he saved. Later he learned the nature of the stones and disposed of them. They were small and brought him from $10 to $50 each.

A man named Schmitz found a diamond in the same locality which weighed one and one-half carats. At French Corral, Nevada county, a number of diamonds have been found, one weighing seven and one-quarter carats. Some fifty or sixty stones have been picked up at Cherokee Flat, the largest of which weighed more than two and one-quarter carats. In a lava-capped channel near Placerville a number of diamonds have been found. Messrs. Jacob Lyon, Thomas Potts, A. Brooks, E. Brentfield, Charles Reed, Jeffries, Thomas Ward, Cruson, and Olmstead are among those who picked up precious stones at that place. The last named man found one which brought him $300. One California diamond is known to have brought $500. A number have sold for $100 each, and several others for half that sum.

In every instance these diamonds have been found in earth or gravel mixed with lava and volcanic ash, and associated with other volcanic matter, such as zircon, platinum, iridium, magnetite, etc. These conditions being known, it is strange that prospectors have neglected to systematically search for this stone. It remains for some prospector to distinguish himself by bringing to light the great diamond field of California.

Mexico offers a field to the prospector in this line, as well as in many others. The following data, published in the Mexican Herald, may serve to point some prospector to unlimited wealth, which would be his if he succeded in opening a prolific field for the production of diamonds. The article relates to the finding of diamonds in the state of Guerrero.

"That General Vicente Guerrero, after whom the state is named, when carrying on a guerrilla warfare at the head of his Indians, during the war of independence from 1810 to 1821, did really find diamonds, and that some of them actually came into the possession of a very highly respected Englishman, the father of the present Archbishop Guillow, and then established in business in the city of Mexico, are facts beyond doubt; but the difficulty has been to discover the exact locality or localities where the General obtained them.

"One authority states that the General found diamonds in the Sierre Madre mountains at a place distant one day and a half by horseback from Tetela del Rio, following the trail to Coronilla. The diamonds were inside hard, egg-shaped stones called by the Indians 'cocos,' which while some were strewn upon the surface, were generally imbedded in hard earth and required a crowbar to dig them out. Breaking the egg-shaped stone or 'cocos' with a hammer or larger stone, the diamond appeared in a hollow in the center, and was aways octahedron or dodecahedron in form, that is, it invariably showed eight or twelve equal faces.

"Another authority states that General Guerrero told him personally that these 'cocos' were also found at a place called Olas de Chilapa, in the district of Chiapa or Tixtla as it is sometimes called."

A variety of gems occur in various parts of the United States and Mexico. The most prolific section, both as to value and variety, is in San Diego county, California. With the exception of the diamond fields of South Africa and Brazil, no other section of country produces so much in value in gems and no section so great a variety or so abundant quantities.

The gem, kunzite, recently discovered by Frank Salmons of San Diego, is thus described by Prof. Kunz, for whom the gem was named:

"A recent remarkable discovery of unaltered lilac-colored spodumene has lately been made in California. The crystals were obtained fifty feet from a deposit of colored tourmaline–itself of notable interest–a mile and a half northeast from Pala, San Diego county. This new discovery is less than a mile northeast ırom the celebrated rubellite and lepidolite locality at that place, where recent developments have brought to light immense quantities of amblygonite, the latter species occuring by the ton, while the lepidolite is estimated by the thousand tons· The locality is thus unequalled in the world for its abundance of lithia minerals.

"The rubellite crystals found here are entirely embedded in lepidolite, the rubellite appearing as radiations of pink in a darker gangue of lilac-covered lepidolite.

"At the new locality colored tourmaline crystals have been found that are remarkable in size and beauty. Although they have been much broken in taking them out, some are a foot long and three inches in diameter with a red crystal core (rubellite) and a blue exterior (indicolite) separated by a pale intervening zone.

"The spoduomene crystals are of extraordinary size, transparency and beauty, and are unrivalled by those from any known localities."

During the year 1904 new gem discoveries of importance were made in several portions of the United States, as follows:

In South Carolina magnificent blue and white topaz, weighing more than a pound apiece, which as crystals quite equalled those of Siberia; beryls from three to six inches long and one or more inches in diameter, varying in color from a pale green to a deep sea-green tint, crystals of rose-colored beryl (a rare variety) found at Mesa Grande, Pala and Hemet; aninite, a gem mineral not known until lately in good crystals in this country, but found in 1904 in beautiful crystals near Ricon; epidote in crystals only one inch long and an eighth of an inch in diameter, but transparent; and magnificent red, green, and yellow tourmalines, found in abundance at Mesa Grande, Pala and other localities in San Diego county.

Near Rumford Falls, Maine, turquoise has been found, and in a number of localities in Arizona, New Mexico and California. The semi-precious stone, peridot, olivine, or chrysolite, has been found in large quantities at Talklai, Arizona, immediately associated with or enclosed in volcanic rock. Thousands of gems one to five carats in weight have been taken out.

The emerald is one of the rarest, most sought, and most valuable of gems today. Its composition is practically the same as beryl, which is not a valuable gem, save in color, and, according to lapidaries, it is a trifle softer. The characteristic beauty of its color, a sparkling velvety green, has never been duplicated by art or matched by nature.

Emeralds, beryls and aquamarines occur in crystals, six-sided prisms, in veins of pegmatite traversing granite, in mica-schists and in gneiss.

Emeralds have been found in limited quantities in North Carolina, New Jersey, in Columbia, South America, in Mexico, and in Egypt. Some of the finest colored specimens come from Peru.

There is always a chance that beryl deposits will yield emeralds, theretore the prospector will do well to thoroughly investigate all beryl deposits.

Tourmaline is not uncommon, but crystals flawless, of good color and sufficiently large to yield good gems are sufficiently rare to give good value to such as as are found. The gem ordinarily occurs in granite, syenite, gneiss, lapidolite, mica-schist, chlorite schist, and occasionally in diorite, granular limestone, etc. The crystals are long, needle-like and become electric when heated.

Corundum, under which species are the ruby, the sapphire and the gems known as Oriental topaz, Oriental amethyst, Oriental emerald, is associated with crystalline rocks, such as granite, gneiss, dolomite, granular limestone, etc. It is usually distinguishable because of its hardness, being next to the diamond. It will itself scratch all other stones, save the diamond and is marked by the diamond alone.

Spinel rubies have usually a dull bluish tinge, though there are green, blue and red varities. These latter have little value except for cabinet specimens.

Topaz has value as a gem. The white topaz resembles, when cut, the diamond, though far less brilliant. The yellow topaz is most in demand as a jewel. The pale blue variety, which occurs in large masses, is used for broaches.

The opal is a favorite stone of those of lesser value, and the true opal, or "noble opal" is classed among the precious stones. This is noted for its fire-like colors, being translucent and emitting the mingled colors of red, yellow, green and blue. The stones are usually found in cavities in lavas, chiefly in amygdaloidal rhyolite.

Turquoise is a pretty blue gem stone and is found in this country, principally in Arizona, New Mexico, and California, though deposits have recently been found in two different localities in Alabama. The gem is opaque and is easily determined by its color, a beautiful sky-blue.

CHAPTER VII.

HOW TO TEST ROCKS AND ORES.

SIMPLE AND INEXPENSIVE TESTS FOR THE PROSPECTOR IN THE FIELD–USE OF THE BLOW-PIPE–ASSAYING IN CAMP–HOW THE RELATIVE VALUE OF ORES MAY BE DETERMINED IN THE FIELD–PREPARING SAMPLES FOR ASSAYING–TIPS THAT SHOULD BE HEEDED.

There are some simple and inexpensive tests which the prospector may apply to the rocks and minerals which he may chance to find, and which will aid him in determining the kind of the mineral before him.

Hardness and streak are two important features in determining the character of the rocks and minerals.

Certain rocks and minerals may be scratched by an ordinary nail, owing to their softness. Among these are gold, silver, sylvanite, nagyagite, petzite, talc, gypsum, calcite, argentite, pyrargyrite, prousite, frieslebenite, stephanite, cerargyrite, or horn silver, embolite, metacinnabarite, native copper, bornite, melaconite, chrysocolla, chalcocite, native lead, galena, anglesite, massicot, minium, pyrolusite, erythrite, stibnite, realgar, orpiment, native bismuth, molybdneum, autunite, torbernite, vanadium.

There are other minerals too hard to be scratched by a nail but still soft enough to be marked by ordinary steel. These are:

Apatite, fluor-spar, platinum, chalcopyrite, cubanite, barnhardite, tetrahedrite, cuprite, malachite, azurite, atacamite, cerussite, pyromorphite, zinc, calamine, zincite, blend, siderite, pyrrhotite, manganese, stannite, genthite, zaratite, cadmium, autunite.

Still harder than these and yet capable of being scratched by the finest tempered steel are:

Franklinite, magnetite, hematite, limonite, ilmenite, chromium, niccolite, gersdorffite, smaltite, cobaltite, turgsten, uranium, orthoclase. Quartz will scratch a few which ordinary steel will not touch and diamond will mar all other minerals.

In testing the hardness of material, care should be taken to find a smooth surface, if possible, for the test, and to use the sharp point or edge of the scratching implement. After the attempt has been made wipe away the dust and examine the piece. You will then detect a scratch upon the surface of the material being tested, or you will note that the implement or mineral used to do the scratching has been dulled.

In connection with this test is one other thing which helps in determining the nature of the mineral in question, that is the "streak." The color of the dust produced in scratching the material, differs in different minerals. This color is called the "streak." It is important that the prospector familiarize himself with the streak of the various minerals. The following list is intended for his help in this line:

Streak white, siderite, calamine.

Silver-white, bismuth, sylvanite, silver.

Tin-white, antimony.

Steel-grey, bournonite, frieslebenite, argenite sylvanite.

Yellowish-grey, calaverite,

Greyish-black, gersdorffite, smaltine, cobaltine, bornite, pyrrhotite, mispickle.

Blackish lead-grey, galena, stibnite, bismuthine, nagygite, argentite, chalcocite,

Black, magnetite, pyrolusite, polybasite, enargite, psilomelane, grey-copper, stromeyerine, copper-glance, metacinnabarite, stannite.

Iron-black, stephanite, polybasite, petzite.

Brown-black, wolfram, nicolite, uranium, braunite marcasite.

Brown, chromium, titanium.

Scarlet, cinnabar, psilomelane.

Cherry-red, hematite, pyrargytite, proustite.

Red-brown, hausmannite, ilmenite, maganese, tungsten, grey copper, manganite, nematite, tennantite, cuprite.

Reddish-bronze, cubanite.

Bronze-yellow, barnhardite.

Metallic-shining, copper.

Pale-green, malachite, zaratite.

Apple-green, atacamite, torbernite.

Blue, azurite.

Yellow, autunite, vanadium.

Orange, minium, arsenic, cadmium.

Lemon, orpiment.

Some minerals, it will be noticed, are listed under two colors. This is because, owing to a variation in the compound of which the ore is composed, they sometimes show one streak and sometimes the other.

The specific gravity of gems has been given in a former chapter. In determining the kind of mineral it is well to know its approximate specific gravity. The prospector, however, will not be provided with the fine scales and other apparatus necessary to determine the exact specific gravity of the materials with which he may meet, therefore, instead of giving a table of exact figures, the minerals have been grouped into three divisions, according to their weight, and the prospector can usually determine, by hefting the stone or mineral, to which group it belongs.

It will be remembered that in fixing the weight or specific gravity of minerals, water is taken as the unit of weight. The specific gravity of gold is 19.5, therefore, bulk for bulk, it weighs nineteen and one-half times heavier than water.

In the first division are placed those minerals which have a specific gravity of less than 3.5, or which are less than three and one-half times heavier than their bulk in water. They are:

Amber, bitumen, lignite, coal, natron, sal-ammoniac, borax, epsom salts, anthracite, potash alum, copperas, saltpeter, sulphate of zinc, sulphur, nitrate of soda, chabazite, graphite, sulphate of copper, rock salt, opal, gypsum, harmotome, quartz, orthoclase, lapsis lazuli, serpentine, beryl, emerald, vivianite, cordierite, albite, anorthite, labrodorite, alunite, talc, steatite, calcite, cryolite, dolomite, magnesite, aragonite, mica, fluor-spar, tourmaline, turquoise, anhydrite, nephrite, apatite, andalusite, calamine, epidote, hornblend, augite.

It will be noticed that most of the soluble minerals are included in this class. The second division contains the minerals whose specific gravity is between 3.5 and 8.5. They are:

Topaz, diamond, olivine, diallogite, realgar, spinel, pleonaste, spathic iron, limonite, strontianite, blend, chrysoberyl, celectine, garnet, rhodonite, azurite, goethite, corundum, malachite, copper pyrites, psilomelane, brookite, rutile, willemite, tin pyrites, chromite, tennanite, manganite, witherite, smithonite, molybdenite, magnetic pyrites, magnetite, tetrahedrite, arsenic, proustite, kerargyrite, hematite, bornite, pyrargyrite, mispickle, cobaltine, cuprite, gersdorffite, anglesite, stephanite, tellurium, pitchblend, cerussite, bismuth glance, tinstone, smaltine, chloanthite, antimony, (metallic), bismuthite, argentite, wolfram, nickeline, galena, cinnabar.

The third division contains those minerals heavier than 8.5. They are:

Gold, platinum, iridium, electrum, mercury, silver, copper, bismuth.

As patterns upon which to base estimations of minerals, it may be remembered that the specific gravity of native, or pure gold, is 19.5, that of pure silver 10.5, native copper 8.84, native lead 11.44, iron 7.8.

Directions have been given, already, for the determining of minerals by color, taste and smell. With the addition of the foregoing information the prospector should be able to arrive at a very correct idea of the minerals he may chance to find.

Let us suppose, for instance, that he has chipped from a ledge with his hammer a bit of rock of a silver-white color. He wishes to determine its character. He hefts the piece in his hand and readily assigns it to the second division of minerals, as given above, for he is quite positive that it is heavier than 3.5 and not so heavy as 8.5.

By thus classing it he has eliminated from consideration all those rocks and minerals in the other two classes. As but few of the minerals in the second division are of a silver-white color, the greater number of them are also out of the question.

He next applies the scratch test and finds that neither the nail nor the steel of his knife will scratch the material, so he obtains a piece of quartz, which has a sharp corner, and with this he makes a scratch upon the surface of the stone. He notes the streak. It is a dark greyish-black. This narrows the matter down to a fine point. Let us see what minerals give this streak. Among those so listed is gersdorffite, but this is a softer mineral and would have been scratched by steel; there is smaltine, but this, also, would have shown the mark of the steel and the color of the mineral is a metallic tin-white, instead of silver-white as is the piece in question. There is cobaltine which is silver-white and gives a similar streak, but that is so soft it would have been scratched by the steel. Bornite also gives the same streak but that is very soft and would have been scratched by the nail and is so light as to be classed in division one as to its weight or specific gravity, and bornite is of a brownish, copper-red color, so this is out of the competition, as is pyrrhotite, which is a bronze-red metal, softer and lighter than our sample. One other mineral is listed as giving the greyish-black streak and that is mispikle. This metal is silver-white, has a gravity of 5.5, hence would be in the second division, would not be scratched by steel, but is softer than quartz. Our sample satisfies all these conditions, so it is fair to presume that it is mispikle.

We have another simple little test, however, which we will apply. If the sample is mispikle, it is strong in arsenic. We take the hammer and strike it sharply and then smell the ore. It gives off the odor of garlic, another of the attributes of mispikle.

This is an illustration of the manner of determining to a very great degree of certainty, the nature of ores without the use of chemicals or blowpipe.

More analytical tests can be made with blowpipe or with the chemicals in camp at the leisure of the prospector. He may, if his camp is so located as to be a permanent headquarters for a considerable time, even introduce a small assaying outfit in camp where he may make very accurate tests, in a small way, of the ores he finds.

The use of the blowpipe and the manner of using the chemical tests has been explained in a previous chapter. In determining the relative value of ore deposits, it will be necessary for the prospector to go down upon the vein to such a point as the ores have not become impoverished by the action of the elements. The width and thickness of the veins should be exposed and samples should be taken for tests from various portions. These samples should be so marked with regards to their position in the vein, that the prospector can tell from what portion of the vein they were taken.

With mortar and chemicals and blowpipe, let him test the samples separately, to begin with, then later let him combine results and make a test of all mixed together.

The first tests are for determining what portions of the vein are mineral bearing, or which parts are richest in minerals, and the latter test is to give him the average run of the vein.

Whenever he has made a find which is shown by his tests to be a promising one, he should prepare a number of samples for the assayer after the manner above described. Only by this method can he arrive at a true knowledge of the value of his claim.

One thing the prospector should bear in mind. That is the relation that rocks and minerals sustain to each other. By noting the presence of certain rocks and minerals one may get the "tip" which will lead to the discovery of rich mineral lodes.

For instance, one may find white mica, fluorspar and apatite in abundance and be little interested because of the trifling value of those minerals but topaz, tourmaline and tin, all valuable, are associated with these minerals and their presence may indicate hidden treasure in those lines. The prospector should be quick to note and heed all of nature's signs.

CHAPTER VIII.

MINING LAWS IN THE UNITED STATES AND MEXICO.

GENERAL LAWS GOVERNING PROSPECTING, LOCATING AND OPERATING IN THE UNITED STATES–MINING DISTRICTS–LOCAL LAWS–SIZE AND EXTENT OF CLAIMS–MINING LAWS IN MEXICO–HOW TO GET A PERMIT TO PROSPECT–SIZE OF CLAIMS AND NUMBER ALLOWED TO THE INDIVIDUAL, ETC.

It is not possible, of course, in the confines of a single chapter, to give all the laws or a digest of all of the laws governing the locating and operating of claims. Such as are most pertinent and important are given, either in full or in substance.

While each state and each mining district has local laws governing these matters, they are not such as interfere or conflict with the United States laws. Regarding local regulations the prospector will have no trouble in informing himself.

Section 2318 of the United States Law says: "In all cases lands valuable for minerals shall be reserved from sale, except as otherwise expressly directed by law.

Section 2319. All valuable mineral deposits in lands belonging to the United States, both surveyed and unsurveyed, are hereby free and open to exploration and purchase, and the lands in which they are found to occupation and purchase, by citizens of the United States and to those who have declared their intention to become such, under regulations provided by law, and according to the local customs or rules of miners in the several mining districts, so far as the same are applicable and not inconsistent with the laws of the United States.

Section 2320. Mining claims upon veins or or lodes of quartz or other rock in place, bearing gold, silver, cinnabar, lead, tin, copper or other valuable deposits heretofore located, shall be governed as to length along the vein or lode by the customs, regulations, and laws in force at the day of their location. A mining claim located after the 10th day of May 1872, whether located by one or more persons, may equal, but shall not exceed 1500 feet in length along the vein or lode; but no location of a mining claim shall be made until the discovery of the vein or lode within the limits of the claim located. No claim shall extend more than 300 feet on each side of the middle of the vein at the surface, nor shall any claim be limited by any mining regulation to less than 25 feet on each side of the middle of the vein at the surface, except where adverse rights existing on the 10th day of May, 1872, render such limitation necessary. The end lines of each claim shall be parallel to each other.

Section 2324. The miners of each mining district may make regulations not in conflict with the laws of the United States, or with the laws of the state or territory in which the district is situated, governing the location, manner of recording, amount of work necessary to hold a mining claim, subject to the following requirements: The location must be distinctly marked on the ground so that its boundaries can be readily traced. All records of mining claims hereafter made shall contain the name of the locators, the date of the location, and such a description of the claim or claims located by reference to some natural object or permanent monument as will identify the claim. On each claim located, until a patent has been issued therefor, not less than $100 worth of labor shall be performed or improvements made during each year. Upon a failure to comply with these conditions, the claim or mine upon which such a failure occurred shall be opened to relocation in the same manner as if no location of the same had ever been made. PROVIDED: That the original locators, their heirs, assigns, or legal representatives, have not resumed work upon the claim after the failure and before the location.

Section 2336. Where two or more veins intersect or cross each other, priority of title shall govern, and such prior location shall be entitled to all ore or mineral contained within the space of intersection; but the subsequent location shall have the right of way through the space of intersection for the purposes of convenient working of the mine. And where two or more veins unite, the oldest or prior location shall take the vein below the point of union, including all the space of intersection.

Section 2335. A patent for any land claimed and located for valuable deposits may be obtained in the following manner: Any person, association, or corporation authorized to locate a claim under this chapter, may file in the proper land office an application for a patent, under oath, showing such compliance, together with a plat and field-notes of the claim or claims, which shall be distinctly marked by monuments on the ground, and shall post a copy of such plat, together with a notice of such application for a patent, in a conspicuous place on the land embraced in such plat previous to the filing of the application for a patent, and shall file an affidavit of at least two persons that such notice has been duly posted, and shall file a copy of the notice in such land office, and shall thereupon be entitled to a patent for the land in the manner following: The register of the land office, upon the filing of such application, plat, field-notes, notices, and affidavits, shall publish a notice that such application has been made, for the period of sixty days, in a newspaper to be by him designated as published nearest to such claim; and he shall also post such notice in his office for the same period. The claimant at the time of filing this application, or at any time thereafter, within the sixty days of application, shall file with the register a certificate of the United States surveyor-general that $500 worth of labor has been expended or improvements made upon the claim by himself or grantors; that the plat is correct, with such further description by such reference to natural objects or permanent monuments as shall identify the claim, and furnish an accurate description, to be incorporated in the patent. At the expiration of the sixty days of publication the claimant shall file his affidavit showing that the plat and notice have been posted in a conspicuous place on the claim during the period of publication. If no adverse claim shall have been filed with the register and the receiver of the proper land office at the expiration of the sixty days of publication, it shall be assumed that the applicant is entitled to a patent, upon the payment to the proper officer of $5 per acre, and that no adverse claim exists; and that thereafter no objection from third parties to the issuance of a patent shall be heard, except it be shown that the applicant has failed to comply with the terms of this chapter.

Section 2329. Claims usually called "placers," including all forms of deposits, excepting veins of quartz, or other rock in place, shall be subject to entry and patent, under like circumstances and conditions, and upon similar proceedings, as are provided for vein or lode claims; but where the land have been previously surveyed by the United States, the entry in its exterior limits shall conform to the legal subdivisions of the public lands.

Section 2330. Legal subdivisions of forty acres may be sub-divided into ten-acre tracts; and two or more persons or associations of persons, having contiguous claims of any size, although such claims may be less than ten acres each, may make joint entry thereof; but no location of a placer-claim, made after the 9th day of July, 1870, shall exceed 160 acres for any one person or association of persons, which locations shall conform to the United States surveys; and nothing in this section shall defeat or impair any any bona fide pre-emption or homestead claim upon agricultural lands, or authorize the sale of the improvements of any bona fide settler to any purchaser.

Section 2333. Where the same person, association, or corporation is in possession of a placer-claim, and also a vein or lode included within the boundaries thereof, application shall be made for a patent for the placer claim, with the statement that it includes such vein or lode, and in such case a patent shall issue for a placer-claim, subject to the provisions of this chapter, including such vein or lode, upon the payment of $5 per acre for such vein or lode claim, and 25 feet of surface on either side thereof. The remainder of the placer-claim not embracing any vein or lode claim, shall be paid for at the rate of $2.50 per acre, together with all costs of proceedings; and where a vein or lode, such as is described in section 2320 is known to exist within the boundaries of a placer-claim, an application for a patent which does not include an application for the vein or lode claim shall be construed as a conclusive declaration that the claimant of the placer claim has no right of possession of the vein or lode claim; but where the existence of a vein or lode in a placer-claim is not known a patent for the placer claim shall convey all valuable mineral and other deposits within the boundaries thereof.

Special laws govern the location and acquisition of title to coal lands and those sections most vitally applying to them are here appended:

Section 2347. Every person above the age of twenty-one years, who is a citizen of the United States, or who has declared his intention to become such, or any association of persons severally qualified as above, shall, upon application to the Register of the proper land office, have the right to enter by legal subdivisions any quantity of vacant coal land of the United States not otherwise appropriated or reserved by competent authority, not exceeding 160 acres to such individual person, or 320 acres to such association, upon payment to the Receiver of not less than $10 per acre, for such lands where the same shall be situated more than fifteen miles from any completed railroad, and not less than $20 per acre for such lands as shall be within fifteen miles of such road.

Section 2348. Any person or associations of persons severally qualified as above provided, who have opened and improved, or shall hereafter open and improve any coal mine or mines upon public lands, and shall be in actual possession of the same, shall be entitled to a preference-right of entry under the preceeding section, of the mines so opened and improved: Provided, That when any association of not less than four persons, severally qualified, as above provided, shall have expended not less than $5000 in working and improving any such mine or mines, such association may enter not exceeding 640 acres including such mining improvements.

According to the Mexican constitution foreigners enjoy the same civil rights in business and mining affairs as citizens of that republic. No affidavit of citizenship is required in obtaining titles to Mexican mining property.

The unit of a mining claim, in Mexico, is a square, one hundred meters to the side, that is, 10,000 square meters, a complete hectare, or a little less than two and one-half acres. This claim continues downward perpendicularly from its four corners and is unaltered in its depth by any change in the vein.

Within a reasonable time after a location is made, usually within sixty days, the prospector in the United States must sink a shaft at least ten feet upon his vein. Thereafter assessment work to the amount of $100 per year must be done. This is not the case in Mexico. All that is required of the prospector is the payment of the annual tax of $10 (Mexican money, equivalent to $5 United States currency)per annum for each "pertenencia" or claim taken up, and he may "denounce," or take up as many claims as he pleases.

A claim carries with it all that is beneath the surface within the boundaries of that claim. Should it be found that a vein located upon a neighboring claim dips and comes under the claim that vein becomes the property of the owner of the claim, under which it has dipped.

Before prospecting for minerals in Mexico, the prospector must go to the magistrate of that district in which he intends to prosecute his search and make declaration of his intentions. He will then be given a permit of that magistrate. Should he wish to carry firearms, he will be obliged to get a permit to do so, else is he liable to arrest and imprisonment if caught with them in his possession.

Labor is cheap in all portions of Mexico. The American prospector can, in consequence, develop his claim very cheaply. Another thing which may interest him is the fact that it is against the law of the county for miners to strike, therefore one need not fear embarassment from that source in operating his mineral claims in that republic.

CHAPTER IX.

HOW TO LOCATE, PROVE AND PROTECT A CLAIM.

WHAT MUST BE DONE IN ORDER TO LOCATE A CLAIM—HOW TO LOCATE—FORM OF NOTICE AND HOW TO STAKE A CLAIM—ASSESSMENT WORK—HOW TO PROVE UP—EXTENT OF CLAIMS—PLACER, LODE AND TUNNEL CLAIMS—MILLSITE CLAIM—WATER RIGHTS—GENERAL INFORMATION.

There are a few things the prospector should bear in mind and one of these is the importance of actually finding mineral before locating a claim. A location made, or claim staked previous to actually finding mineral thereon is void.

When some startling discovery has been made and there is a rush for the lucky territory, it frequently happens that hundreds of acres of territory adjacent to the "strike" is staked and notices posted without a shovelful of earth having been disturbed or a particle of mineral having been discovered on the lands so staked. The locations are made on the chance that mineral will be found there, or from the belief that a claim so close to the valuable find will sell for a large sum. Such claims are invalid. They are no claims at all. It is the actual discoverer of mineral who is entitled to locate.

If gold, platinum, tin, or other metal, gems or soluble minerals are found in any quantity mixed with the soil, sand or gravel, a placer claim may be located without first proving to what extent the mineral exists or the actual value of the deposits.

If a quartz vein, or vein, lode or dike of any material be found to contain minerals, it may likewise be located without waiting to prove the value of the same, though the width, thickness and direction of the vein should be, if possible, ascertained.

For the guidance of the prospector some extracts from the Land Office Regulations are here given. They are sufficiently explicit to serve him in proceeding to locate any minerals he may discover:

Any person who is a citizen of the United States, or who has declared his intention to become a citizen, may locate, record, and hold a mining claim of 1500 linear feet along the course of any mineral vein or lode subject to location; or an association of persons, severally qualified as above, may make a joint location of such claim of 1500 feet, but in no event can a location of a vein or lode exceed 1500 feet along the course thereof, whatever may be the number of persons composing the association.

With regard to the surface-ground adjoining a vein or lode, and claimed for the convenient working thereof, the Revised Statues provide that the lateral extent of the location of veins or lodes shall in no case exceed 300 feet on each side of the middle of the vein at the surface, and that no such surface rights shall be limited by any mining regulations to less than 25 feet on each side of the middle of the vein at the surface, the end lines of such claims to be, in all cases parallel to each other. Such lateral measurements cannot extend beyond 300 feet on either side of the middle of the vein at the surface, or such distance as is allowed by the local laws. For example: 400 feet cannot be taken on one side and 200 feet on the other. If, however, 300 feet on either side are allowed, and by reason of prior claims but 100 feet can be taken on one side, the locator will not be restricted to less than 300 feet on the other side; and when the locator does not determine by exploration where the middle of the vein at the surface is, his discovery shaft must be assumed to mark such point.

By the foregoing it will be perceived that no lode-claim can exceed a parellelogram 1500 feet in length by 600 feet in width, but whether surface-ground of that width can be taken depends upon the local regulations of state and territorial laws in force in the several mining districts; and that no such local regulations of state and territorial laws shall limit a vein to less than 1500 feet along the course thereof, whether the location be made by one or more persons, nor can surface rights be limited to less than 50 feet in width.

It is provided by the Revised Statutes that the miners of each district may make rules and regulations not in conflict with the laws of the United States nor with the state or territory in which such districts are respectively situated, governing the location, manner of recording, and amount of work necessary to hold a claim. They likewise requirə that the location shall be so distinctly marked on the ground that its boundaries may be readily traced. This is a very important matter and locators cannot exercise too much care in defining their locations at the outset, inasmuch as the law requires that all records of mining locations shall contain the names of the locators, the date of the location, and with a description of the claim or claims located, by reference to some natural object or permanent monument as will identify the claim.

The statutes provide that no lode claim shall be recorded until after the discovery of a vein or lode within the limits of the claim located, the object of which provision is evidently to prevent the appropriation of presumed mineral ground for speculative purposes to the exclusion of bona fide prospectors, before sufficient work has been done to determine whether a vein or lode really exists.

The claimant should, therefore, prior to locating his claim, unless the vein can be traced upon the surface, sink a shaft, or run a tunnel or drift, to a sufficient depth therein to discover and develop a mineral-bearing vein, lode, or crevice; should determine, if possible, the general course of the vein in either direction from the point of dicsovery, by which direction he will be governed in marking the boundaries of his claim on the surface. His location notice should give the course and distance as nearly as practicable from the discovery shaft on the claim, to some permanent, well-known points,or objects, such, for instance, as stone monuments, blazed trees, the confluence of streams, points of well-known gulches, ravines or roads, prominent buttes, hills, etc., which may be in the immediate vicinity, and which will serve to perpetuate and fix the locus of the claim and render it susceptible of identification from the description thereof given in the record of locations in the district, and should be duly recorded.

In addition to the foregoing data, the claimant should drive a post or erect a monument of stones at each corner of his surface-ground, and at the point of discovery or discovery shaft should fix a post, stake, or board, upon which should be designated the name of the lode, the name or names of the locators, the number of feet claimed, and in which direction from the point of discovery; it being essential that the location notice filed for record, in addition to the foregoing description, should state whether the entire claim of 1500 feet is taken on one side of the point of discovery, or whether it is partly upon one one side and partly upon the other side thereof, and in the latter case, how many feet are claimed on either side of such discovery point.

Within a reasonable time, say twenty days after the location shall have been marked on the ground, or such time as is allowed by the local laws, notice thereof, accurately describing the claim in the manner aforesaid, shall be filed for record with the proper recorder of the district, who will thereupon issue the usual certificate of location.

In order to hold the possessory right to a location not less than $100 worth of labor must be performed, or improvements made thereon annually until such entry shall have been made. Under the provisions of the act of Congress approved January 22nd, 1880, the first annual expenditure becomes due and must be performed during the calendar year succeeding that in which the location was made. Expenditure made of labor performed prior to the first day of January suceeding the date of location will not be considered as a part of, or applied upon the first annual expenditure required by law. Failure to make the expenditure or perform the labor required will subject the claim to relocation by any other party having the necessary qualifications, unless the original locator, his heirs, or asigns, or legal representatives, have resumed work thereon after such failure and before such relocation.

The expenditures upon mining claims may be made from the surface or in running a tunnel for the development of such claims, the act of February 11th, 1875, providing that where a person or company has, or may, run a tunnel for the purpose of developing a lode or lodes owned by said person or company, the money so expended in said tunnel shall be taken and considered as expended upon such lode or lodes, and such person or company shall not be required to perform work on the surface of such lode or lodes in order to hold the same.

The importance of holding to these details in the matter of location, labor, and expenditure will be more readily perceived when it is understood that to give the subject improper attention may invalidate the claim.

In posting a notice of location of a claim the prospector should always, if it is possible to do so, have a witness to the posting of the notice and the witness should attest the notice. The following is a legal form for the location notice:

Notice of Location.

Notice is hereby given that the undersigned, having complied with the requirements of Chapter Six of Title Thirty-two of the Revised Statues of the United States, and the local customs and laws and regulations, has located (here state whether placer, lode or tunnel claim and the extent of the claim) situated in....................Mining District.. .... County ....
State, and described as follows: (Here give a complete description as per the foregoing instructions)
Discovered..........................
Located..........................
Witnesses: ....Locator.
....
....

The proving up of a claim, making final entry and obtaining a patent is accompanied by a considerable detail and again the instructions of the Land Office are given for the benefit of the prospector.

The claimant for patent is required in the first place, to have a correct survey of his claim made under the authority of the surveyor-general of the state or territory in which the claim lies; such survey to show with accuracy the exterior surface boundaries of the claim, which boundaries are required to be distinctly marked by monuments on the ground. Four plats and one copy of the original fields notes, in each case, will be prepared by the surveyor-general, one plat to be given the claimant for posting upon the claim, one plat and the original field notes to be retained in the office of the surveyor-general, one plat and a copy of the field notes to be given the claimant for filing with the proper register, to be finally transmitted by that officer, with other papers in the case, to the Land Office, and one plat to be sent by the surveyor-general to the register of the proper land district to be retained on his files for future reference.

The claimant is then required to post a copy of the plat of such survey in a conspicuous place upon the claim, together with notice of his intention to apply for a patent therefor, which notice will give the date of posting, the name of the claimant, the name of the claim, mine or lode; the mining district and county; whether the location is of record, and, if so, where the record may be found; the number of feet claimed along the vein, and the presumed direction thereof; the number of feet claimed on the lode in each direction from the point of discovery or other well-defined place on the claim; the name or names of adjoining claimants on the same or other lodes; or, if none adjoin, the names of the nearest claims, etc.

After posting the said plat and notice upon the premises, the claimant will file with the proper register and receiver a copy of such plat, and the field notes of survey of the claim, accompanied by the affidavit of at least two creditable witnesses, that such plat and notice are posted conspicuously upon the claim, giving the date and place of such posting; a copy of the notice so posted to be attached to, and to form a part of, said affidavit.

Accompanying the field notes so filed must be the sworn statement of the claimant, that he has the possessory right to the premises therein described in virtue of a compliance by himself (and by his grantors, if he claims by purchase) with the mining rules, regulations, and customs of the mining district, state or territory in which the claim lies, and with the mining laws of Congress; such sworn statement to narrate briefly, but as clearly as possible, the facts constituting such compliance, the origin of his possession, and the basis of his claim to a patent.

This affidavit should be supported by appropriate evidence from the mining recorder's office as to his possessory right, as follows: Where he claims to be the locator, or a locator in company with others who have since conveyed their interest in the location, to him, a full, true and correct copy of such location should be furnished, as the same appears upon the mining records; such copy to be attested by the seal of the recorder, or, if he has no seal, then he should make oath to the same being correct, as shown by his records. When the applicant claims only as a purchaser for valuable consideration, a copy of the location record must be filed under seal or upon oath, as aforesaid, with an abstract of title from the proper recorder, under seal or oath, as aforesaid, brought down as near as practicable to the date of filing the application, tracing the right of possession by a continuous chain of conveyances from the original locators to the applicant, also certifying that no conveyances affecting the title to the claim in question appear of record in his office other than those set forth in the accompanying abstract.

In the event of the mining records in any case having been destroyed by fire or otherwise lost, affidavit of the fact should be made and secondary evidence of the possessory title will be received, which may consist of the affidavit of the claimant, supported by those of any other parties cognizant to the facts relative to his location, occupancy, possession, improvements, etc.; and in such case of lost records, any deeds, certificates of location or purchase, or other evidence which may be in the claimant's possession and tend to establish his claim, should be filed.

Upon the receipt of these papers the register will, at the expense of the claimant, publish a notice of such application for the period of sixty days in a newspaper published nearest to the claim, and will post a copy of such notice in his office for the same period. When the notice is published in a weekly newspaper ten consecutive insertions are necessary; when in a daily newspaper the notice must appear in each issue for sixty-one consecutive issues, the first day of issue being excluded in estimating the period of sixty days.

The notices so published and posted must be as full and complete as posible, and embrace all the data given in the notice posted upon the claim.

Too much care cannot be exercised in the preparation of these notices, inasmuch as upon their accuracy and completeness will depend, in a great measure, the regularity and validity of the whole proceedings.

The claimant, either at the time of filing these papers with the register or at any time during the sixty days of publication, is required to file a certificate of the surveyor-general that not less than $500 worth of labor has been expended or improvements made upon the claim by the applicant or his grantors; that the plat filed by the claimant is correct; that the field notes of the survey, as filed, furnish such an accurate description of the claim as will, if incorporated into the patent, serve to fully identify the premises, and that such reference is made therein to natural objects or permanent monuments as will perpetuate and fix the locus thereof.

It will be the more convenient way to have this certificate indorsed by the surveyor-general, both upon the plat and field-notes of survey by the claimant as aforesaid.

After the sixty days' period of newspaper publication has expired, the claimant will furnish from the office of publication a sworn statement that the notice was published for the statutory period, giving the first and last day of publication, and his own affidavit showing that the plat and notice aforesaid remained conspicuously posted upon the claim sought to be patented during the said sixty days' publication, giving the dates.

Upon the filing of this affidavit, the register will, if no adverse claim was filed during the period of publication, permit the claimant to pay for the land according to the area given in the plat and field-notes of the survey aforesaid, at the rate of $5 for each acre and $5 for each fractional part of an acre, the receiver issuing the usual duplicate certificate therefor. The claimant will also make a sworn statement of all charges and fees paid by him for publication and surveys, together with all fees and money paid the register and receiver of the land office; after which the whole matter will be forwarded to the Commissioner of the General Land Office and a patent issued thereon if found regular.

Section 2323 of the United States Law provides that where a tunnel is run for the development of a vein or lode, or for the discovery of mines, the owner of such tunnel shall have the right of possession of all veins or lodes within three thousand feet from the face of such tunnel on the line thereof, not previously known to exist, discovered in such tunnel, to the same extent as if discovered upon the surface; and locations on the line of such tunnel of veins or lodes not appearing upon the surface, made by other parties after the commencement of the tunnel, and while the same is being prosecuted with reasonable diligence, shall be invalid; but failure to prosecute the work on the tunnel for six months shall be considered an abandonment of the right to all undiscovered veins or lodes on the line of said tunnel.

The effect of this is simply to give the proprietor of a mining tunnel run in good faith the possesory right to fifteen hundred feet of any blind lodes cut, discovered, or intersected by such tunnel, which were not previously known to exist, within three thousand feet from the face or point of commencement of the tunnel, and to prohibit other parties, after the commencement of the tunnel, from prospecting for and making locations of lodes on the line thereof and within said distance of three thousand feet, unless such lodes appear upon the surface or were previously known to exist. The three thousand feet are measured from the point where the tunnel actually enters cover.

The procedure for the obtaining patents to tunnel and placer claims are in a general way the same as in obtaining patent for a lode claim, with the exception that in the case of placer claims upon surveyed lands no further survey or plat will be required. All placer claims must conform as nearly as practicable with the United States system of public land surveys and the rectangular subdivisions of such surveys, and no such location shall include more than twenty acres for each individual claimant. Also in paying for the lands, at the time of taking the patent, but $2.50 per acre is required for placer claims instead of $5 per acre as for lode claims.

Where valuable lodes of minerals are found it is sometimes of utmost importance to their development that a suitable site for a mill be obtained. The law provides that such a site may be taken in a manner similar to the taking of a mineral claim. The section applying to millsites is appended:

Section 2337. Where non-mineral land not contiguous to the vein or lode is used or occupied by the proprietor of such vein or lode for mining or milling purposes, such non-adjacent surface-ground may be embraced and included in an application for a patent for such vein or lode, and the same may be patented therewith, subject to the preliminary requirements as to survey and notice as are applicable to veins or lodes; but no location hereafter made of such non-adjacent land shall exceed five acres, and payment for the same must be made at the same rate fixed by this chapter for the superficies of the lode. The owner of a quartz-mill or reduction works, not owning a mine in connection therewith, may also receive a patent for his mill site, as provided in this section.

One of the important things, in developing a mine or claim, whether it be placer or lode, is water. After locating mineral in paying quantities and staking the claim, the prospector should turn his attention to the water supply. If no stream is on or near his claim he should prospect the neighborhood for some stream which may be so situated that its waters can be converted by flume or canal to his claim and if any such be found he should at once file a water claim by posting a notice near the outlet of the stream and recording the same in the district or county recorder's office. If other claims to water in the stream are recorded ahead this does not prevent his filing a claim for as many inches as he may require, only he will not be allowed to take his water from the stream until the former claimants have what they are entitled to, provided they demand their full proportion.

The following form of notice will prove sufficient:

Notice of Right to Water.

The undersigned claims the water running in this........................stream to the extent of ....inches for mining purposes, to be conveyed by (ditch, flume, or canal) from this point to the ....claim.

Dated.............................

........................Locator.

General information regarding mining regulations, local laws, local customs and any points which may arise not covered by the foregoing, may be obtained at the nearest land office if the prospector is unable to gain the information of other prospectors or miners in the locality.

CHAPTER X.

MINING MISCELLANY.

General hints and miscellaneous information—Some mistakes that are made—Prospecting at home—A few final words—Glossary of terms peculiar to mining and prospecting and Dictionary of Minerals.

Before bringing this work to a close a few general hints to the prospector may not be amiss.

In searching for minerals he should not overlook the fact that certain rocks and earths are valuable, as well as certain deposits not generally reckoned among the minerals, and that fortunes are frequently made in locating these unusual materials.

Marbles, white and colored, are always valuable if in sufficient mass and reasonably accessible. Building, monumental and statuary granite is valuable under favorable circumstances. Salt deposits, rock salt, the soluble minerals all, are usually worth the prospector's attention. Petroleum oil and natural gas are most valuable, and are often found through seepages which are manifest upon the surface of the earth.

Lithographic stone, a very fine-grained, compact limestone, free from veins and fossils, is very valuable. The stone should be of light color and capable of being cut in large blocks or slabs.

Magnesite, used for the manufacture of Epsom salts, is usualy found in talcose schists, serpentine, or other magnesian rocks, and is valuable if found in abundant quantities.

Meerscham, which may be recognized by its smooth compact texture, its lightness, and the fact that when dry it will adhere to the tongue, is also valuable. Its streak is white and slightly shining. It is usualy found in serpentine rocks in nodular masses. With a cobalt solution it gives off a pink color before the blowpipe.

Mica, if found in large sheets, is valuable. Lapidolite, or lithia mica, is used in making a mineral water known as lithia water. It is valuable if found in quantity and reasonably near a railroad and where it can be mined reasonably easy.

Some clays are very valuable. Kaolin, or porcelain clay, plastic or potters' clay, and fullers' earth, a clay used for freeing wool from fatty matter, are valuable if conditions are right for the mining and shipping.

All are liable to mistakes, and the prospector is not an exception to the rule. One of the mistakes he is liable to, especially if he be a novice, is overhaste and lack of thoroughness, or, perhaps overeagerness would better describe this fault. He wants to hurry from place to place, feeling that the great prize is somewhere ahead of him. This feeling leads him to neglect the territory over which he passes, giving it but a superficial examination, as he flits.

The superficial prospector seldom meets with success. It is the deliberate, patient, delving, analytical prospector who picks the plums. Running about is not prospecting. Prospecting is getting down to the roots of things and seeking diligently.

If he camps at the foot of a mountain, the character of the exposed rocks of which leads to a reasonable supposition that the mountian is the home of valuable minerals, the prospector should not leave the vicinity till he is reasonably assured that such is not the case, or till he has located and secured a claim upon such minerals. If there are promising signs his time may as well be spent there as in running about frolicking with Chance.

Another mistake to which the prospector is prone is self-deception. He wants to find rich deposits of mineral. When he does find a lode or placer containing mineral he is quite apt to let his desire influence his estimate of the value thereof. However much satisfaction this may afford him at the time, it only leads to future loss and disappointment. He should be very judicial and impartial in forming his conclusions in these matters.

Another mistake frequently made is the attempting to locate too many claims. The average prospector is a poor speculator. There are too many folks with gold mines and mineral claims for sale, to make it profitable to bring to market any but those of proven value. The prospector must remember that to hold a claim or to make his location valid, he must operate and improve it. If it isn't a good claim it isn't worth the work necessary to hold it. If it is a good claim, he better put in his time developing it than in seeking to locate other claims. Too many claims are the prospector's undoing.

One other mistake may be referred to just here, and that is carelessness in making his location and proofs. If a claim proves to be a particularly valuable one there are plenty of unscrupulous persons who are ready to take advantage of any slip, oversight or technical negligence on the part of the discoverer, and who may file contesting claims and attempt to deprive him of his property. Even if they should not succeed in their efforts, it is expensive and unpleasant to have to enter into litigation over a claim. Great care should be exercised in marking out the claim, setting substantial monuments, posting all required notices in a proper manner and fulfilling every demand of the law, in order to leave no grounds for litigation.

In the glossary of terms and definitions and explanations relative to minerals, earths and rocks, which forms the bulk of this chapter, it is hoped that the prospector may find such additional information as he may require in prosecuting his search for nature's hidden valuables.

Let it be said in closing, that those would-be prospectors who do not have the time or equipment for lengthy trips or protracted periods of investigation, may, by a careful perusal of this book and by practical application of its teachings, become actual prospectors at home.

Few sections of country but contain minerals of some sort that are valuable. Few localities but still have their mineral secrets. Almost every day we read of valuable discoveries being made in the very midst of populous centers. We are a blind race, even if we do possess eyes, and there are yet plenty of opportunities in most localities. If one cannot be a roaming prospector, he may, therefore, be a home prospector.

And whether he prosecute his search in familiar and well-known fields near home, or scales the rugged heights of distant mountains, or yet traverses the weary wastes of desert sands, may success ever attend the prospector is the wish of the writer.

GLOSSARY AND MINERAL DICTIONARY.

Acid................A salt of hydrogen in which the hydrogen may be replaced by a metal.

Actinolite...........A species of hornblende.

Adamantine.........Resembling a diamond.

Adularia............A species of orthoclase, composed of alumina and potash with amethyst, jasper or quartz.

Agate..............Silicate, chalcedony alternated with amethyst, jasper or quartz.

Alabandine..........Sulphide of manganese.

Alabaster...........A compact form of gypsum.

Albite..............Felspar, composed of alumina and soda.

Allophane...........A hardened clay composed of hydrous silicate of alumina.

Alluvial.............Material washed together by water.

Almandine..........Noble garnet.

Alum...............Sulphate of alumnia with potash, magnesia, soda, amonia, or iron.

Alumnia............Oxide of aluminum.

Aluminum..........One of the metallic elements.

Alunite............. Sulphate alumnia and potash.

Amalgam........... A silver ore; silver combined with mercury.

Amalgamation........ The absorbing of gold by mercury.

Amber.............. Fossil gum.

Amethyst........... A silicate gem or purple quartz.

Amorphous.......... Not crystallized; without specific form.

Amphibolite .. Hornblende rock.

Amygdaloids ..... Small almond-shaped cavities in igneous rocks containing minerals.

Anancime .... Hydrous silicate of alumina and soda.

Anamesite .... Basalt of medium texture.

Andalusite .... A silicate of alumina.

Andesite............ An igneous rock of glassy appearance, varying in composition.

Anglesite............ Sulphate of lead.

Anhydrite........... Sulphate of lime.

Anhydrous.......... Having no water in its composition.

Annabergite......... Arseniate of nickle.

Anorthite........... A kind of felspar composed of lime and silicate of alumnia.

Anthracite.......... Non-bituminous coal.

Anticline............ Strata bent like the letter A; a saddleback.

Antimonides........Combinations of antimony with metals.
Antimony...........One of the metallic elements.
Apatite.............A phosphate of lime.
Apopyhlite..........Hydrous silicate of lime with potassium fluoride.
Aquamarine.........A variety of beryl; silicate of alumina and glucina.
Aqueous Rocks......Rocks the material of which has been deposited in water, either by chemical precipitation or as a sediment.
Aragonite...........Rhombic carbonate of lime.
Argentiferous.........Silver-bearing.
Argentite...........Sulphide of silver.
Arkose.............Rock formed by disintegrated granite.
Arquerite...........Silver ore consisting of silver and mercury.
Arseniates..........Compounds of arsenic acid with bases.
Arsenic.............A metallic element.
Arsenides...........Compound of arsenic with metals.
Asbestos............A variety of hornblend composed of long flexible filaments.
Asphaltum..........Mineral resin.
Augite..............A silicate of lime, magnesia and iron.
Auriferous..........Gold-bearing.

Azurite..............Blue carbon of copper.
Bacillar..............Gathered in bundles, like faggots.
Backs..............Ground between a level in a mine and the next level above.
Banded veins.. ....Veins made up of layers of different minerals running parallel with the walls of the vein.
Banket..............Gold-bearing conglomerates cemented with quartz.
Barytes..............Sulphate of baryta.
Barytocalcite........Carbonate of baryta and lime.
Basalt..............A volcanic rock.
Bases..............Compounds which are converted into salts by the action of acids.
Bedded Veins.......Beds which conform to the trend of the strata in which they occur, both in strike and dip.
Bedding............The arrangement of strata in layers.
Bed Rock...........Solid or stratified rock underlying soil and alluvial deposits.
Beryl...............One of the gems; a silicate of alumina and glucina.
Biotite.............Magnesian iron mica.
Bismuth............One of the metallic elements.

Bismuth Glance.....Sulphide of bismuth.
Bismuthine.........Same as above.
Bismuth Ochre......Oxide of bismuth.
Bismuthite..........Hydrous carbonate of bismuth
Bitumen............Mineral pitch or tar.
Black Jack..........Zinc blende.
Blende.............Sulphide of zinc.
Bloodstone..........A variety of quartz, dark-green with specks of veins of jasper.
Blowpipe...........A metal pipe for forcing air through flame to increase and apply the heat to some object.
Boracite............A borate and chloride of magnesia.
Borax..............A borate of soda.
Bornite.............Sulphide of copper and iron.
Botryoidal..........In a cluster, as a bunch of grapes.
Bournonite..........Sulphide of lead, antimony and copper.
Braunite............Anhydrous oxide of manganese.
Brazilian Emerald...A kind of tourmaline.
Brazilian Sapphire...Tourmaline resembling sapphire in color.
Brecca..............A conglomerate of angular fragments.
Breithauptite........Antimonide of nickle.

Brine .... Water containing salt in solution.

Brilliant .... Applied to surfaces that are perfect reflectors of light; the diamond is a "brilliant."

Brittle .... Breaking easily.

Bromargyrite......... Bromide of silver.

Bromide .... A compound of bromine with a metal.

Bromine .... An element.

Bronzite .... A silicate of magnesia and iron.

Brookite .... A form of oxide of titanium.

Brown Coal .... A hydrous coal.

Brown Iron Ore ..... A hydrous oxide of iron.

Bunches .... Detatched irregular masses of ore in a vein.

Cadmium .... One of the metallic elements.

Cairngorm .... Smoky quartz.

Calamine .... Carbonate of zinc

Calaverite .... Telluride of gold.

Calcareous .... Containing carbonate of lime.

Calcareous Sinter ... Calcareous deposit from water.

Calceraous Tufa ..... Same as above.

Calcination ... Roasting at a gentle heat.

Calcite ...... Hexagonal carbonate of lime.

Calcium .... A metallic element.

Calomel ..... Chloride of mercury.

Calmei .... A silicate of iron.

Cannel Coal .... Coal rich in illuminating gas.

Capillary .... Hair-like.

Carat .... $3\frac{1}{6}$ troy grains.
Carbonaceous........ Containing fossil carbon.
Carbonas........... Irregular offshoots of mineral from lodes.
Carbonates .... Compounds of carbonic acid with a base.
Carbonic Acid .... Carbonic anhydride.
Carnallite .... Chloride of magnesia and potash.
Carnelian .... A red species of quartz.
Casing .... Clayey material found between a vein and its walls.
Cassiterite.......... Oxide of tin.
Cat's Eye .... True cat's eye is chrysoberyl; false cat's eye is quartz enclosing asbestos fibers.
Caunter Lode ...... A diagonal lode.
Celestein ...... Sulphate of strontia.
Cement .... Auriferous conglomerate.
Cerussite ..... Carbonate of lead.
Cervantite .... Oxide of antimony.
Chabazite ...... Hydrous silicate of alumina, lime, potash and soda.
Chalcanthite........ Sulphate of copper.
Chalcedony......... A kind of quartz.
Chalcopyrite........ Sulphide of copper and iron.
Chiastolite .... Silicate of alumina.
Chloanthite .... Arsenide of nickle.
Chlorides.......... Compounds of chlorine with metals.

Chlorination ..A process of converting gold, in separating it from other metals, into chloride of gold by action of chlorine.

Chlorine ....An element.

Chlorite .....A hydrous silicate of magnesia and alumina.

Chlorobromides .Compounds of chlorine and bromine with metals.

Chrome Iron........A chromate of iron.

Chromite...........The same.

Chromium ....One of the metallic elements.

Chrysoberyl .....A gem composed of aluminate of glucina.

Chrysocolla.........A hydrous silicate of copper.

Chrysolite ....Silicate of magnesia and iron.

Chrysoprase ....An apple-green variety of quartz.

Chrysotile ....Hydrous silicate of magnesia.

Cinnabar ....Sulphide of mercury.

Citrine ......Yellow quartz; false topaz.

Clay ....A hydrated silicate of alumina minutely divided.

Clay Slate ...Slate formed by the induration of clay.

Cleavage ......Property possessed by certain rocks and minerals of splitting in certain directions.

Cleavage Planes.....The plane along which cleavage takes place.

Coal ....Fossilized carbon composed of vegetable matter.

Cobalt ....A metallic element.

Cobalt Bloom.......Arseniate of cobalt.

Cobaltiferous Wad .Impure oxide of manganese containing cobalt.

Cobaltine ....Sulphide and arsenic of cobalt

Columar ....Having the forms of columns.

Combustible ..Capable of being burned.

Compact ......Close-grained; not showing special crystals.

Conchoidal .....Resembling the inside of a clam shell in shape.

Concretion ....Nodule of mineral formed around some object or center.

Congenial ....Rocks in which the veins have become ore-bearing.

Conglomerate Consolidated gravel or rounded pebbles.

Contact Deposits Mineral deposits occuring at the junction of two dissimilar rocks.

Contortion ....Crumpling or twisting of strata or veins.

Copper ....One of the metallic elements.

Copperas ....Sulphate of iron.

Copper Glance .Sulphide of copper.

Copper Nickle ....Arsenide of nickle.

Copper Pyrites......Sulphide of copper and iron.

Copper Slate .... Slate impregnated with copper.
Coprolites........... Phosphate of lime.
Cordierite........... Silicate of alumina, iron and magnesia. A gem.
Corundum.......... A gem alumina.
Coveline............ Sulphide of copper.
Cradle.............. Box mounted on rockers used in washing gold from gravel.
Craterform.......... In the form of a crater.
Crevicing........... Searching crevices in rocks for gold deposits.
Crocidolite.......... A fibrous silicate of iron, soda and magnesia.
Crocoisite........... Chromate of lead.
Cross Course........ A vein cutting across an older vein.
Cryolite............. Fuoride of alumnia and soda.
Crystallization....... The taking, by minerals, of definite geometrical forms.
Cube............... A solid six-sided figure of which each of its sides is a square.
Cuprite............. Red oxide of copper.
Cyanidation......... Converting gold into a double cyanide of potassium and gold by action of potassium.
Cyanite............. A silicate of alumina
Decrepitate......... Bursting into fragments before the blowpipe.
Degradation......... Wearing away.

Dehydrated.........Deprived of water.
Dendritic ....Like branches of a tree.
Denudation.........Stripping as by water.
Detritus............Accumulation from disintregated rocks surfaces.
Diallage............A silicate of lime and magnesia
Diallogite...........Carbonate of manganse.
Diamond...........Crystallized carbon. A precious stone.
Diatoms............Tiny plants having a siliceous covering.
Dichroic............Exhibiting two different colors when light is transmitted in different directions.
Dichroite...........Corderite; a silicate of alumina, iron and magnesia.
Diopside............A silicate of lime and magnesia
Dioptase............A silicate of copper.
Diorite.............A volcanic rock.
Dip................Angle of inclination of strata.
Disintregation......The crumbling of solid matter owing to the action of chemical or physical forces.
Dislocation.........Displacement of rock on either side of a crack.
Disthene............Cyanite; a silicate of alumina.
Dolerite............An igneous rock.
Dolomite...........Carbonate of lime and magnesia.
Domes.............Strata dipping away in all directions.

Domekite............Arsenide of copper.
Dredging............Lifting material from below water by a machine called a dredge.
Drift...............Loose matter displaced by water.
Drusy...............Cavities in rocks lined with crystals.
Dry Ores............Silver ore which does not contain lead.
Ductile.............Capable of being drawn into a wire.
Dunite..............A massive olivine rock interspersed with fine grains of chromate.
Dyke or Dike........Veins filled with porphry or similar rock.
Efflorescence........Crystals or powder appearing on the surface of minerals due to their decomposition.
Elastic.............Substances which may be stretched but which will resume their original form.
Elaterite............Elastic bitumen.
Electrum............An alloy of silver and gold.
Elements............Substances never decomposed or incapable of analysis.
Embolite............Chlorobromide of silver.
Emerald.............A gem composed of alumina and glucina.

Emery............An impure variety of corundum.

Enargite............Sulphide, arsenide and antimonide of copper.

Enstatite...........A silicate of magnesia and iron.

Epidote.............Hydrous silicate of alumina, iron and lime.

Epsom Salt.........Hydrous sulphate of magnesia.

Eroding............A wearing away.

Erubescite..........Sulphide of copper and iron.

Eruptive............A breaking out of enclosed matter.

Erythrine...........Arseniate of cobalt.

Excrescence.........Grown out of.

Exfoliate...........To peel off in layers from the outside.

Fahlbands..........Zones of crystalline schists impregnated with metallic substances which contribute to veins which pass through them.

Fault...............Displacement of strata caused by fracture of the earth's crust.

Felspars............Anhydrous silicates of alumina and alkali or lime.

Ferruginous.........Containing iron.

Fibrous.............Composed of fibres which are not easily separated.

Filiform............Thread-like, but not massed together as when fibrous.

Fire Clay...........A silicate of alumina nearly free from lime or alkalies, capable of standing a high degree of heat.

Fissures............Open cracks in the rocks.

Flat................A fracture which is smooth like a board.

Flats of Ore.........Horizontal ore deposit.

Flocculent..........Resembling lumps of wool.

Floor...............Bottom of a coal seam.

Flucan..............A soft, clayey casing.

Fluor Spar..........Fluoride of lime.

Foliated............Made up of thin leaves, like a book.

Footwall............The lower side of a lode.

Fossiliferous.........Rocks containing fossils.

Franklitein..........Oxide of iron, zinc and manganese.

Free Milling.........Ores which yield their gold or silver without chemical treatment.

Freislebenite........A sulphide of silver, lead and antimony.

Fullers' Earth.......Soft unctuous clays used in treating woolen goods.

Gabbro..............A volcanic rock.

Galena..............Sulphide of lead.

Galmei..............A silicate of zinc.

Gangue.............Matrix in which mineral occurs.

Garnets...............Anhydrous silicates of alumina and earths colored by oxides of iron, manganese, and chromium.

Garnierite...........A silicate of nickle.

Gash Veins..........Fissures confined to certain rocks and not extending to abutting rocks.

Ge-Anticlinical......Dome-shaped bendings of the earth's crust.

Gelatinise...........Become like jelly.

Geo-Synclinal.......The reverse of Ge-Anticlinal.

Gersdorffite.........An arsenide of nickle.

Geysers.............Intermittant boiling springs.

Glacial Deposits.....Deposits made by glaciers during the glacial period.

Glauber Salt........Sulphate of Soda.

Glaucodote..........A kind of cobaltine.

Gneiss..............Stratified granitoid rock in which the minerals are in layers.

Goethite............A hydrous oxide of iron.

Gold...............One of the metallic elements.

Goslarite............Sulphate of zinc.

Gossan.............Hydrated peroxide of iron, often quartzose, found capping veins containing ferruginous minerals.

Granite.............An igneous rock.

Granular............In the form of grains.
Graphite.............A form of carbon.
Greisen..............Granite composed of mica and quartz.
Grey Copper........Tetrahedrite, a copper ore.
Grossularia..........Green garnet.
Gypsum.............Hydrous sulphate of lime.
Haematite..........Anhydrous oxide of iron.
Halloysite...........A hard clay.
Hanging Wall........The upperside boundary of a lode or vein.
Harmotome ....Hydrous silicate of alumina and baryta.
Hauerite........... Sulphide of manganese.
Hausmannite.......Anhydrous oxide of manganese.
Hauyne............Silicate of alumina, soda, and sulphate of lime.
Heave.............A lateral displacement of a lode produced by a fault.
Hedenbergite.......Black augite.
Heliotrope..,........Bloodstone–dark-green quartz specked or veined with jasper.
Hessite ....Telluride of gold.
Heulandite.........Hydrous silicate of alumina and lime.
Homogeneous ......Having the same structure throughout.
Hornblende..........A silicate of lime, magnesia, and iron.

Horn Silver.........Chloride of silver.

Hyacinth...........A variety of zircon; a gem.

Hydraulic Lime.....Lime that will set or become solid under water.

Hydrocarbons ......Compounds of carbon and hydrogen.

Hydromagnesite ..Hydrous carbonate of magnesia.

Hydrothermal......The action of hot water in dissolving and precipitating minerals in the earth.

Hydrous............Containing water of crystallization.

Hypersthene ....A silicate of magnesia and iron.

Hypersthenite.......Rock formed of labradorite and hypersthene.

Iceland Spar ....Crystallized transparent carbonate of lime.

Idocrase...........A silicate of alumina, lime and magnesia.

Igneous............That which has been transformed by subterranean heat or fires.

Impregnation........Ore disseminated through rocks having no well-defined limits.

Indicator Vein.........A non-metalliferous vein which leads to ore deposits.

Indurated..........Hardened.

Infusorial Earth ....A siliceous deposit composed principally of diatoms.
In Situ ....In the place where formed.
Interstratified ....Interbedded with.
Intrusion ....Forcing through.
Intumescence........Swelling when heated.
Iodargyrite..........Iodide of silver.
Iodine..............A metallic element.
Iridescent..........Displaying many colors, like a rainbow.
Iridium............One of the metallic elements.
Iridosmine ....An alloy of iridium and osmium.
Iron ....Ametallic element.
Iron Pyrites........Sulphide of iron.
Isomorphism........The property possessed by certain chemical compounds of replacing one another in minerals.
Itacolumite ....Flexible sandstone.
Jade...............A silicate of lime, magnesia and iron.
Jamesonite.........A sulphide of lead and antimony.
Jargon.............A kind of zircon.
Jet ....A hard kind of coal cut and polished for ornaments.
Jigger..............A machine for the concentration of ores.
Kaolin..............A pure clay.
Kerargyrite ....Chloride of silver.

Kerosene Shale......Shale containing illuminating oil.

Killas ....Clay slate.

Labradorite.........A silicate of lime, alumina, and soda.

Laccolites ....Lenticular sheets of eruptive rock injected between beds.

Lameliar...........In thin sheets.

Laminae ....Thin plates or scales.

Lapis Lazuli.........Ultramarine.

Laumontite ....A silicate of lime and alumina.

Lava ....Rock which has flowed from volcanoes in a molten state.

Leaders.............Small offshoot veins from lodes containing the same kind of minerals.

Leads ....The gold-bearing portions of alluvial deposits which mark the former course of the stream.

Lenticular..........Like a lens.

Lepidolite..........A lithia mica.

Lherzolite..........Pyroxene-olivine rock.

Libethenite.........Phosphate of copper.

Lignite ....A hydrous coal retaining its woody structure.

Lime ....Oxide of calcium produced by calcining carbonate of lime.

Limestone..........Rock formed of carbonate of lime.

Limonite............Hydrous oxide of iron.

Linnaeite............Sulphide of nickle and cobalt.
Lithographic Stone ..Very fine-grained limestone.
Lode ....Any vein which carries metals.
Magma ....Paste or groundwork of igneous rocks.
Magnesia............Oxide of magnesium.
Magnesite..........Carbonate of magnesia.
Magnetic Pyrites ....Pyrrhotine; a sulphide of iron.
Magn etite..........Magnetic oxide of iron.
Malachite............Green carbonate of copper.
Malleable............Capable of being moulded.
Mammillary.........In smooth, round prominences.
Manganese..........A metallic element.
Manganite ....Hydrous oxide of manganese.
Marble..............Metamorphic limestone.
Marcasite............Rhombic sulphide of iron.
Massive ....Not crystallized.
Matrix..............The rock or mineral containing metallic ores or gems.
Meerchaum.........Hydrous silicate of magnesia.
Melaconite..........Black oxide of copper.
Melanite............Black garnet.
Melaphyre ....An igneous rock.
Mercury.............One of the metallic elements.
Metallic..............Resembling polished steel.
Metalliferous.........Metal-bearing.
Metamorphism......Change in mineral composition and internal structure of rocks brought about by heat, pressure, etc.

Micaceous..........Made up of thin plates like flakes of mica.

Micas...............Flexible minerals occuring in thin plates, composed of alumina and potash, magnesia, lithia, or iron.

Mica Schist.........A metamorphic rock consisting of laminated quartz and mica.

Millerite.............Sulphide of nickle.

Mimetite............Arseniate of lead.

Minium..............Oxide of lead.

Mispickle ....Sulphide and arsenide of iron.

Molybdenite ....Sulphide of molybdenum.

Molybdenum ....A metallic element.

Molybdite...........Oxide of molybdenum.

Moonstone ....A variety of adularia felspar.

Moraines............Deposits made by glaciers.

Mountain Leather ...An impure variety of absetos.

Mundic..............Pyrites; sulphide of iron.

Muscovite..........Potash mica.

Nacreous...........Resembling mother of pearl.

Nagyagite...........Telluride of gold and lead.

Natrolite............Hydrous silicate of alumina and soda.

Natron ....Carbonate of soda.

Nephrite ......Jade.

Nickle ....One of the metallic elements.

Nickeline ....Arsenide of nickel.

Nickel Ochre ....An arseniate of nickel.

Nicopyrite ....Sulphide of nickel and iron.

Nitrates ....Compounds of nitric with bases.
Nitratine............Nitrate of soda.
Nodular ....Concretions of rock matter aggregated around a central nucleus.
Noumeite............Silicate of nickel.
Nucleus ....A body about which anything is collected.
Obsidian............A volcanic glass.
Octahedrite.........Titanic oxide.
Octahedron ....A figure having eight equal sides.
Ondontolite.........False turquoise; fossil bone colored by copper.
Oligiste.............An oxide of iron.
Oligoclase...........Silicate of alumina, soda and lime.
Olivine...............Silicate of magnesia and iron.
Onyx...............A quartz in alternate layers of white and brown or white and black.
Oolitic..............Made up of minute rounded particles.
Opal ....Hydrous silica.
Opaque.............Not permitting the passage of light.
Opalescence.........Exhibiting a play of colors similar to the opal.
Oriental Amethyst A variety of corundum.
Oriental Emerald.....A variety of corundum.

Oriental Topaz .....A variety of corundum.
Orpiment...........Sulphide of arsenic.
Orthoclase ....Silicate of alumina and potash.
Outcrop ....Appearance at the surface of a rock, lode or coal seam.
Oxides.............Compounds of oxygen with any element.
Ozokerite...........Mineral wax.
Palladium...........A metallic element.
Pan Amalgamation...Combining gold or silver with mercury by grinding in a pan.
Partings............Bands of shale or slate occur ing in a seam of coal.
Peacock Ore........Copper pyrites which has become tarnished.
Pearly...............Resembling mother of pearl.
Pegmatite..........Veins of coarsely crystallized granite running through granite.
Pennine............A kind of chlorite.
Perlite.............A volcanic glass.
Petrified...........Changed to stone.
Petroleum..........A bituminous mineral oil.
Petrology...........The study of rocks.
Petzite.............Telluride of silver and gold.
Phenakite...........A gem composed of silicate of glucina.
Phosphate...........Compounds of phosphoric acid with a base.

Pipes of Ore.........Elongated bodies of ore standing nearly or quite perpendicular.

Pisolitic Ores........Ores in concretions about the size of a pea.

Pitchblende..........Oxide of uranium.

Pitchstone...........Volcanic glass.

Plasma..............Green quartz.

Plastic..............Easily moulded.

Platiniridum.........An alloy of platinum and iridium.

Platinum............One of the metallic elements.

Pleonaste...........A kind of spinel.

Plications...........The smaller foldings of a rock.

Plumbago...........Graphite.

Plolybasite..........A sulphide of silver, copper, antimony, and arsenic.

Porphrite...........Porphry containing crystals of plagioclase.

Porphry............Any igneous rock consisting of a ground mass containing crystals.

Potash.............Oxide of potassium.

Precipitate..........Solids in solution deposited by the addition of another solution.

Prehnite............Silicate of alumina and lime.

Prismatic...........In prisms.

Prisms.............Solids whose bases are plane figures and whose sides are parallelograms.

Productive.......... Yielding minerals in paying quantities.

Propylite........... Felspar and hornblende ground together.

Proustite........... Sulphide and arsenide of silver.

Psilomelane......... Manganate of baryta.

Pumice............. A vesicular volcanic glass.

Purple of Cassius.... Precipitate obtained from adding stannous chloride to chloride of gold.

Pyrargyrite......... Sulphide of antimonde of silver.

Pyrites............ Cubic sulphide of iron.

Pyrolusite.......... Black oxide of manganese.

Pyromorphite....... Phosphate of lead.

Pyrope............. A species of garnet.

Pyrosmalite......... Silicate of iron and magnesia.

Pyrrhotine.......... Magnetic pyrites.

Quarried........... Worked in the open.

Quartz............. Crystallized silica.

Quartz Diorite ...... An igneous rock.

Quartzite........... Metamorphic sandstone.

Quicklime.......... Oxide of calcium made by burning limestone.

Quicksilver......... Mercury.

Radiating.......... Diverging from a common center.

Ramified............ Branched in many directions.

Realgar............. Sulphide of arsenic.

Red Lead............Oxide of lead; minium.
Redruthite..........Copper glance.
Reduction...........Reducing compounds to a metallic state.
Reefs...............Mineral ledges.
Refraction..........Deviation from a direct course.
Refractory..........Refractory ores are those difficult to treat for recovery of metals therein.
Reniform............Kidney-shaped.
Reticulated Veins....Veins that run in all directions.
Reverse Faults......Faults in which the hanging wall has been thrust upwards on the footwall.
Rhodonite...........Silicate-of manganese.
Rhombic Dodecahedron..A twelve-sided figure each side of which is a romb.
Rhombohedron......A six-sided figure, each side of which is a romb.
Rhyolite.............An igneous rock.
Ribbon Veins........Same as banded veins.
Rim Rock...........Bed rock which outcrops at a higher level than a placer bed.
Ripidolite............A kind of chlorite.
Rock Crystal........Clear, colorless quartz.
Rock Salt...........Chloride of Sodium.
Roof................The strata next above a coal seam.
Rubellite...........Red tourmaline.

Ruby................Red corundum.
Rutile..............A form of oxide of titanium.
Saccharoid..........Like lump sugar.
Saddle Reefs........Quartz reefs that bend in saddle-shape.
Sal-Ammoniac.......Chloride of ammonium.
Saltpeter............Nitrate of potash.
Sandstone...........Consolidated sand.
Sandine..............A glassy variety of orthoclase.
Sapphire............Blue corundum.
Sardonyx............A kind of quartz.
Scarp................A steep face.
Scheelite ....Tungstate of lime.
Schist ....Laminated metaphoric rock.
Schorl..............Black tourmaline.
Sectile ....Capable of being cut with a knife.
Section ....A division; a cut through.
Sediment ....Deposit formed by water.
Sedimentary........Rocks composed of sediment.
Segregations.........Collections of ores in cavities, irregular in form and defined as to limits.
Serpentine ....Hydrous silicate of magnesia.
Selenium............An element.
Shale..............Consolidated clay.
Shingle.............Clean gravel.
Shining ....Opposite of dull or earthy, as describing the streak in metals.

Shoots . . . . Deposits of ore in lodes whose principal extent is downward, or at an angle.
Siderite . . . . Carbonate of iron.
Silica . . . . Oxide of silicon.
Silicates . . . . Compounds of silica or silicic acid with a base.
Silver . . . . One of the metallic elements.
Silver Glance . . . . Sulphide of silver.
Sinter . . . . A deposit from hot springs.
Slags . . . . Fusible silicates formed by smelting ores.
Slates . . . . Indurated clays.
Slide . . . . A fault.
Slimes . . . . Fine particles of crushed ores that float on top of water.
Sludge Channel . . . . . . Race for carrying tailngs away from placer workings.
Sluice Box . . . . Trough for washing gold from gravel.
Smaltine . . . . Arsenide of cobalt.
Soapstone . . . . Compact variety of talc.
Soda . . . . Oxide of sodium.
Spathic Iron . . . . Carbonate of iron.
Spherosiderite . . . . Carbonate of iron.
Spinel . . . . Aluminate of magnesia.
Stalactites . . . . Incrustations hanging down from the roof of caves.
Stalagmites . . . . Incrustations extending upwards from cavern floors.
Stannine . . . . Sulphide of tin and copper.

Steatite............. Hydrous silicate of magneous.
Stephenite.......... Sulphide and antimonide of silver.
Sternbergite......... Sulphide of silver and iron.
Stibnite............. Sulphide of antimony.
Stockworks.......... Rock interspersed with numerous mineral veins.
Stratification........ Arrangement of sedimentary rocks in layers.
Streak.............. Powder of a mineral produced by scratching it with a harder substance.
Striated............ Marked with furrows.
Strike.............. Horizontal line upon the floor or footwall of a lode.
Stromeyerine........ Sulphide of silver and copper.
Strontia............ Oxide of strontium.
Sulphates........... Compounds of sulphuric acid with a base.
Sulphides........... Compounds of sulphur with metals.
Sulphur............ An element.
Sulphuretted Hydrogen.. A sulphide of hydrogen.
Sunstone........... A kind of oligoclase.
Syenite............ An igneous rock.
Sylvanite.......... Telluride of gold and silver.
Syncline........... Strata bent in the form of a trough.
Tachylyte.......... A volcanic glass.
Tailings............ Refuse from a mine or mill of

ore from which the mineral has been extracted.

Tail Race........... Channel for carrying away the tailings.

Talc................. Hydrous silicate of magnesia.

Tellurium........... One of the metallic elements.

Tennantite.......... Sulphide and arsenide of copper and iron.

Tetrahedrite......... A complex copper ore.

Thomsonite.......... Hydrous silicate of alumina, lime and soda.

Till.................. A glacial deposit.

Timazite............ Hornblende andesite.

Tin Pyrites.......... Sulphide of copper and tin.

Tinstone............ Oxide of tin; cassiterite.

Titanates........... Compounds of titanic acid with a base.

Titanic Iron......... Specular iron with oxide of titanium.

Toad's Eye Tin ..... Wood tin in small round parcles embedded in a mass of lighter or darker shade.

Topaz............... A gem composed of silicate of alumina with fluorine.

Torbanite.......... A dark brown variety of cannel coal.

Tourmaline......... A gem composed of silicate of alumina and other oxides.

Trachyte............ An igneous rock.

Translucent......... Transmitting light but not transparent.

Transparent........Transmitting light perfectly.
Trappean...........Rocks occuring in dikes and sheets.
Travertine ....Material deposited by calcareous springs.
Tremolite...........White hornblende.
Triclinic ....Crystals which have three axes which are not at right angles.
Tridmyte...........Silica occuring in small hexagonal tables.
Tungatates......... Compounds of tungstic acid with a base.
Tungsten...........A metallic element.
Ultramarine........Lapis lazuli.
Underclay..........Clay bed of coal seams.
Underlay...........Inclination of lodes to the vertical.
Uranium ....One of the metallic elements.
Uwarowite..........A chrome garnet.
Valentinite ....Oxide of antimony.
Vesuvianite ....Silicate of alumina, lime and magnesia.
Vitreous............Glassy.
Vivianite...........Hydrous phosphate of iron.
Volcanic............That which came from a volcano.
Wad...............Impure earthy ore of manganese.
Waxy..............Resembling the surface of clean beeswax.

Willemite............Anhydrous silicate of zinc.
Witherite............Carbonate of baryta.
Wolfram............Tungstate of iron and manganese.
Woolastonite........Silicate of lime.
Wood Tin ....Tinstone of a brown color reniform in shape and fibrous in structure.
Wulfenite ....Molybdate of lead.
Zeolites ....Hydrous silicates of alkalies, or alkaline earths with silicates of alumina.
Zinc ....A metallic element.
Zinc Blende ....Sulphide of zinc.
Zinc Bloom ....Hydrous carbonate of zinc.
Zincite.............Red oxide of zinc.
Zinkenite ....Sulphide of lead and antimony
Zircon ....A silicate of zirconia; a gem.
Zoiste ....A non-ferriferous rhomboic ally of epidote.

INDEX